Signals and Images

Leoni Villano Bonamin
Editor

Signals and Images

Contributions and Contradictions about High Dilution Research

Springer

Editor
Prof. Leoni Villano Bonamin
Universidade Paulista
São Paulo
Brazil

ISBN 978-1-4020-8534-5 e-ISBN 978-1-4020-8535-2

Library of Congress Control Number: 2008928515

Printed on acid-free paper

9 8 7 6 5 4 3 2 1

springer.com

Preface

The "Groupe International de Recherche sur l'Infinitésimal" (GIRI) was created in 1986 by Professor Madeleine Bastide and Doctor René Halm. Madeleine Bastide was its President for several years. The group comprises actually more than 100 researchers from different disciplines (physicians, chemists, biologists, pharmacologists, physicists, etc.) from 22 different countries. The original aim, which, for the most part, has now been achieved, was to create a working group to exchange ideas and results concerning the effects of very low doses and high dilutions. Madeleine Bastide then proposed to publish a collection of the lectures given at their scientific meetings in a bilingual book (French and English) entitled "Signals and Images".

The first volume was published in 1990 by the ATELIERS ALPHA BLEUE. It contains lectures on the new pharmacological approach and related concepts that were presented at the 3rd and 4th Symposiums held in Paris in 1989 and 1990. This volume treats the problem of the effects of very low doses and high dilution in *in vitro* and *in vivo* experimental models. It is conceivable that this effect, like that of electromagnetic fields, is none other than an "ultra molecular" effect as shown in yet published results. The effect could be interpreted as a piece of information, i.e. a signal whose transmission and perception remains to be elucidated. According to Madeleine Bastide "everything depends on the nature and quality of the information, signals and images". Among possible concepts, it was proposed to go beyond the mechanistic paradigm (which has enabled many outstanding discoveries) and to envisage the science of positive interactions or *Sense-paradigm* that is "one of organized matter whose determining reality is information". One potential fruitful approach is to envisage that these two paradigms exist side by side and to examine them both within their own fields and the possible interconnections between them.

The second volume was published in 1997 by Kluwer Academic Publishers. It contains the synthesis of the main lectures at the 7th and 8th Symposium of the GIRI held in Montpellier and Jerusalem in 1993 and 1994. The lectures are grouped in four chapters:

- Hormesis: a concept that is often wrongly used to explain the law of similarity.
- *In vitro* and *in vivo* models: a report on the effects of high dilution of various substances based on well-documented experiments.

- Therapeutics and proving: an analysis of symptoms through a range of therapeutic trials on humans and animals.
- Epistemological approach: an other approach that considers the living organism not as an object, but as a complex structure, i.e. a locus of communications that is continuously modified as a function of time and enables exchange of information with the internal and external environment and the processing of such information. This approach led Madeleine Bastide and Agnès Lagache to propose the *Paradigm of signifiers* as a way to improve our understanding of the effects of high dilution in homeopathy and to interpret them.

This third volume is a compilation of selected findings presented between the 16th Symposium and the 21st Symposium of the GIRI, including the one held in São Paulo in 2006, the first meeting organized in the American continent; which gave new colors to the group. A scientific committee, composed by Dr. Michel Van Wassenhoven; Dr. Catherine Gaucher; Dr. Richard Blostin; Prof. Carlos Renato Zacharias and Prof. Leoni Villano Bonamin did the selection of lectures. All chapters were submitted to peer-review before to be published. Researchers observed the bioethical aspects according their own countries statements.

The publication of the 3rd volume demonstrates the will of the GIRI members to continue the work that Madeleine Bastide pursued with courage, perseverance and dedication to strengthen homeopathy through rigorously managed high-level research, conducted by an international multidisciplinary group of competent researchers.

The present volume was structured in a similar format then the formers. It is divided in six parts: epistemology, in which the concepts proposed by Madeleine Bastide and Agnès Lagache are discussed upon new experimental results, basic research – physics, basic research – biology, clinical research, veterinary research and practice and an epilogue, in which some reflections about the interest of homeopathy of promoting health in developing countries are exposed by Dr. Catherine Gaucher, who founded the "Homeopathes Sans Frontières".

As a natural result of different and multidisciplinary approaches, some chapters can be apparently contradictories, some concepts that are discussed can be quite polemic, some results can be unexpected.... But this is science. This is the GIRI's face. Undoubtedly, high dilutions, homeopathy and related disciplines are still opened subjects.

Dear reader, enjoy the book! It is our most sincere wish.

Professor Jean-Marie Bastide
Professor Leoni Villano Bonamin

Contents

Preface v

Part I Epistemology

1 **Research on Ultra-dilutions and the Theory of Corporeal Signifiers: The Follow Up** 3
Leoni Villano Bonamin, Agnès Lagache, and Madeleine Bastide

Part II Basic Research: Physics

2 **Theoretical and Physical-Chemical Models for Dynamized Systems: Validation Criteria** 29
Carlos Renato Zacharias

3 **Mechanical Versus Handmade Succussions: A Physical Chemistry Comparison** 37
Carla Holandino, Felipe Dias Leal, Bianca de Oliveira Barcellos, Maria Augusta Campos, Raíza Oliveira, Venício Feo da Veiga, Sheila Garcia, and Carlos Renato Zacharias

4 **Water and High Dilutions Phenomenology: Physical Characterization** 49
Adriana Ramos de Miranda

Part III Basic Research: Biology

5 **Non-linearity Modeling of Ultra-dilutions: The Histamine Disturbances Case** 67
Guadalupe Ruiz-Vega and Gabino Estevez-Delgado

6 **Hepatic Cell Growth Models for the Study of Ultra High Dilutions** 83
Kátia Silva Martinho, Vanessa Del Bianco de Bento, Graziella Ulbricht Benvenga, Viviane Aparecida Marcondes, and Leoni Villano Bonamin

7 **Models with Plants, Microorganisms and Viruses for Basic Research in Homeopathy** 97
Lucietta Betti, Grazia Trebbi, Daniele Nani, Vera Majewsky, Claudia Scherr, Tim Jäger, and Stephan Baumgartner

8 ***Arnica montana* and Behavior of Connective Tissue** 113
Leoni Villano Bonamin

9 **How Living Organisms Detect Very Low Doses? A New Approach to Homeopathy and Hormesis** 127
Elżbieta Malarczyk

10 **The Effect of Homeopathically Prepared Thyroxin (10^{-30} Parts by Weight) on Highland Frogs: Influence of Electromagnetic Fields** 137
Sabrina Weber, Peter Christian Endler, Sonja U. Welles, Erika Suanjak-Traidl, Waltraud Scherer-Pongratz, Michael Frass, Heinz Sprange, Gerhard Peithner, and Harald Lothaller

Part IV Clinical Research

11 **Clinical Verification of Homeopathic Symptoms: Experience and Statistics** 151
Lex Rutten and Michel van Wassenhoven

Part V Veterinary Research and Practice

12 **Homeopathic Veterinary Clinical Protocol** 185
Luiz Figueira Pinto

Epilogue

13 **Some Reflections About the Interest of Non-conventional Medicine in Public Health, Especially Homeopathy** 205
Catherine Gaucher

Index 213

Part I
Epistemology

Chapter 1
Research on Ultra-dilutions and the Theory of Corporeal Signifiers: The Follow Up

Leoni Villano Bonamin[1,2], Agnès Lagache[3], and Madeleine Bastide[4] (in memoriam)

Introduction: Science and Homeopathy

More than 200 years ago, when Homeopathy was first conceived by Samuel Hahnemann, the scientific view was being consolidated within a mechanistic paradigm, whose lines had been drawn in the 17th century. That is to say, it was the task of science to know the material world from factual evidence, through observations and conclusions *a posteriori*, which included the formulation or the understanding of a *mechanism*. Still in our days, it may be said that for an explanation to be considered scientific, among other requirements it must posit a *mechanism* for one or more related phenomena (Maturana, 1997).

According to Hahnemann, as seen in his writings, particularly the *Organon of the Healing Art*, the explanations on the mechanisms of diseases suggested by his contemporaries still carried the weight of the arbitrariness and authoritarianism prevalent in the previous centuries, when the phenomena were frequently interpreted following poorly justified inferences (Hahnemann, 1996). On disputing such arbitrariness, Hahnemann was consistent with the *intellectual milieu* of his time. Being so, the question must be asked: why did Hahnemann choose to construct Homeopathy on doctrinaire grounds, rather than as a medical discipline? Why was Homeopathy marginalized from the scientific universe, in a time when modern medicine was being institutionalized? Perhaps it was precisely the "encapsulation of the homeopathic knowledge" in a *doctrine* format, which allowed its ideas survive for more than two centuries without loosing their vitalistic roots. Similarly, by assuming this format, Homeopathy would be protected from the questioning of its

[1]Universidade Paulista, São Paulo, Brazil. leonibonamin@gmail.com

[2]Universidade de Santo Amaro, São Paulo, Brazil

[3]Licée Carnot, Paris, France

[4]Univeristé Montpellier I, Montpellier, France

Presented part at the XXth Symposium of GIRI, São Paulo, 2006 and part at the XXIth Symposium of GIRI, Monaco, 2007

L.V. Bonamin (ed.) *Signals and Images*,

mechanisms, which would have been unavoidable if exposed to the mechanistic view. This would have led to its immediate and obvious degradation. Even now, in the 21st century, the mechanism of action of the homeopathic preparations remains completely unknown; on the other hand, a chronic process of self-exclusion of the homeopathic community prevails somehow attached to anachronical notions, regarding the antinomy "vitalism *vs.* mechanicism" (Priven, 2005).

The development of mechanistic science from the 18th century on was explosive. A wide technological development followed, which perhaps would not have been possible if science would have remained under the limits of the religious authority, as it was during the middle ages (Atlan, 2007). However, the attempt to explain the world according to linear causality, characteristic of the mechanical view of classic science, is unable to reach categories, which take into account the whole and its reciprocal relationships (Amaral, 2007); it means that the mechanistic scientific explanations are unable to solve all the problems. Recently, vitalism has come back to the center of the stage, but in such a poorly structured way that it opens the door to deceitful speculations. On the other hand, the contemporary conception of the living beings and their properties grounded on the notions of semiosis and autopoiesis is an emergent reality (Maturana, 1997; Barbieri, 2007). Perhaps, this new direction may allow devising methods suitable to the demonstration of the *similia principle* not as a dogma, but as a simple biological phenomenon, available to systematic observation. In short: neither vitalism nor mechanicism, but both at one and the same time. A third paradigm, grounded on these two roots apparently opposed, may create a proper intellectual environment for a new understanding of the world, science, and – why not? – Homeopathy.

Experimental Data Involving Ultra-dilutions: The Problem of Interpretation

The *similia principle* (similar is healed by similar) is understood as the correspondence between the symptoms manifested spontaneously by a patient and the symptoms elicited by a substance according to its toxicological and pathogenetic aspects. *Ultra-dilutions* or *high dilutions* refer to the extremely low molarity preparations of biologically active substances, frequently above the number of Avogadro. In the case of the homeopathic remedies, the ultra-dilutions are prepared through serial dilution followed by agitation, a process called *potentization* or dynamization. For this reason, the ultra-dilutions prepared according to the homeopathic pharmacotechnics may also be called *dynamized systems*.

In the 1980's, a new movement emerged within the homeopathic universe. European researchers, not necessarily homeopaths but interested on the subject, conducted experimental studies aiming to demonstrate, with full methodological rigor, the *similia principle* and the effects of ultra-dilutions, the two essential pillars of the homeopathic doctrine and, at the same time, its most polemical features. In 1986 it was founded the first, and only, scientific society exclusively devoted to the study of ultra-dilutions: the *Groupe International de Recherche sur l'Infinitésimal* – GIRI, through an initiative by Prof. Madeleine Bastide, at the time, Chair of the Department of Immunology at *Université de Montpellier I* and by Dr. René Halm, pharmacist from Nice and the head organizer of *Les Entretiens Internationaux de Monaco*. The *Theory of the Corporeal Signifiers*, the central object of this chapter, is one or the most discussed subjects in the GIRI annual meetings.

In 1988, the most polemical paper on the subject was made public by the journal *Nature*. It was authored by Jacques Benveniste and his team, and employed a classic experimental model: basophile degranulation (Davenas et al., 1988). Ultra-dilutions of anti-IgG were used to elicit the degranulation of basophiles previously *in vitro* sensitized by IgE. The oscillatory curve obtained was considered a positive result and it was attributed to the so-called "memory of water", a notion evoked even in the present time as the supposedly scientific explanation of Homeopathy. However, other researchers could not reproduce this result at the time (Hirst et al., 1994; Wiegant, 1994).

Some years later, members of the team, now leaded by Philippe Belon, continued the studies using this same model with some structural modifications: the inhibition of sensitized basophiles degranulation was elicited through the incubation of Ig-E sensitized cells with ultra-diluted *histamine*. This model was exhaustively reproduced, also in a multicentric study involving six universities in different countries (Belon et al., 2004). In their latest publication on this subject, Sainte-Laudy and Belon (2006) showed, after 47 repetitions, that the expression of CD 203c – a specific marker for degranulated basophiles – presented its peak of inhibition in the presence of histamine 7CH and 17CH, while it was absent in the presence of histamine 12CH, which reinforces the idea of an oscillatory potency-effect curve. The most interesting finding in this study was the specificity of the phenomenon: the incubation of methyl-histamine 17CH and histidine 17CH simultaneously with anti-IgE elicited no effect at all. Thus, there are two limiting factors: the dilution above the threshold for the observation of molecular phenomena and the *specific,* putative *modulator* nature of the ultra-diluted histamine activity.

Apparently contradictory, the data above show an intrinsic difficulty to the study of ultra-dilutions. Each new research brings a new surprise. The results appear almost as if random, no prediction can be made about them. Naturally, the first impulse is to attribute such results to technical artifacts. However, this is, in fact, a complex issue, which requires paying closer attention.

Upon analyzing both studies, it is perceived that, although the model employed was the same – *in vitro* degranulation of basophiles – the experimental design was very different. In the second study (Belon et al., 2004; Sainte-Laudy and Belon, 2006), the demonstration of the phenomenon is unquestionable but the model was grounded on the notion of **modulation**, instead of **triggering** of the biological effect, as it was aimed to establish in 1988 (Davenas et al., 1988). This reveals a putative very important property of the phenomenon: its modulator character.

Among the biochemical phenomena, especially those related to inflammatory mediators, the same molecule is able to play a direct or a modulator effect on a cellular system, according to the context and the presence or not of other simultaneous factors.

Histamine itself may act as a direct agent on histaminergic receptors or as modulator of the cells that released it, be it a basophile, a mast cell or a neuron (Bonamin and Abel, 2002). Nevertheless, in the case of ultra-dilutions, the effects seem to be exclusively modulator and may be observed in experimental models not necessarily related to the inflammatory process (Bastide et al., 1987; Daurat et al., 1988; Youbicier-Simo et al., 1997; Bonamin et al., 2001; Coelho et al., 2006). Thus, the properties emerging from the experimental observations, on the one hand supply hints on the nature of the phenomenon and, on the other, lead us away from the known biochemical mechanisms. If the observations of experimental data continue, maybe other properties could be revealed and added to these ones. Thus, it would be possible to build a kind of *mosaic* that, together with a compatible rational basis, may be a body of ideas which will allow make predictions – as can be easily done in classical pharmacology – and, hence, formulate a scientific explanation.

Experimental Illustrations

In 2002, Bonamin and Martinho showed some particularities of the effects of ultra-diluted dexametasone on hepatic proliferation (Bonamin et al., 2002). Rats subjected to partial hepatectomy regenerate the hepatic tissue in about 30 days. A simultaneous treatment with dexametasone 7CH (10^{-17}M) elicits no changes in the recovery of the functional parenchyma. However, a simultaneous treatment with dexametasone 4 mg/kg delays this recovery, indifferent to its association or not with dexametasone 7CH. As the pharmacological effects of dexametasone on the inhibition of cell proliferation are well known, this experiment *a priori* did not bring any new data. Nevertheless, the application of the same experimental model on hepatectomy submitted rats that were submitted to carcinogenesis induction with initiator drugs (acetylaminofluorene) resulted in a very different outcome. It was observed that the animals treated with dexametasone 7CH, associated or not with dexametasone 4 mg/kg, presented a larger number of preneoplasic nodes 30 days after hepatectomy. Similarly, the phenotype of these nodes was more aggressive. More details about this study are seen in Chapter 6.

Thus, another peculiarity of ultra-dilutions was been revealed: their biological activity was highly **dependent on the context**. The modulator effects of dexametasone 7CH on the growth and phenotype of proliferating cells could only be evidenced under pathological conditions, involving a huge recruitment of local stem-cells (oval cells) and an intense activation of proto-oncogenes, differently that occurs under normal conditions of hepatic regeneration, when mainly the remaining hepatocytes multiply, guided by the basal membrane. This observation also refers to another important feature: the **sensitiveness of certain stem cells** to the ultra-dilutions.

This finding corroborates classic studies on this subject developed in the 90's. Youbicier-Simo et al. (1996, 1997) showed that the treatment of *in ovo* bursectomized chickens with ultra-diluted bursine (10^{-27}) elicited the recovery of the specific humoral immune reaction in levels comparable to the control animals. This did not happen when ultra-dilutions of unspecific peptides of similar molecular weight were employed. Endler et al. (1995) showed in a multicentric study that the addition of ultra-dilutions of thyroxine in the water of basins containing tadpoles of *Rana temporaria* delayed metamorphosis, under high altitude conditions. Guedes et al. (2004) reproduced this result under tropical low altitude conditions on *Rana catesbeiana*. This series of data shows that the processes of cell growth and differentiation may be fitting models for the study of ultra-dilutions and may contain important clues regarding their mechanisms.

It must be emphasized that all these examples are related to experimental models in which endogenous or analogous substances are processed according to the homeopathic pharmacotechnics and employed as modulator factors of biological processes related to the original physiological functions. These models are called **iso-endopathy**, as they do not strictly mirror the *similia principle*, which is exclusively established on the analogy of symptoms. Thus, it is important to take into account the possibility to exist more than one mechanism of action of ultra-dilutions.

In some *in vitro* studies, cells exposed to cytotoxic agents and previously treated with isotherapic preparations of the same agents presented higher viability than the controls. However, the pre-treatment of cells with ultra-dilutions of a different cytotoxic agent from the one employed to intoxicate them did not elicit the same effect (Delbancut et al., 1997). In the same manner, variations in the cell pattern may also trigger different results. Walchli et al. (2006) showed that very high potencies of isotherapic of cadmium protected only the cells obtained from primary cultures, and did not protect cells originated in the tumoral lineage.

From these isopathic models two important features may be grasped: the **specificity** of the protection and the need of some degree of **cell organization**, in order to allow the manifestation of the protection effect. That is to say, the manifestation the effect described above depended on the previous conditions of the involved living system, contrary to the observation of classic toxicology, where the toxic effects are directly related to the nature of the molecules and their interactions with the biological membranes.

This dependence on the previous conditions of the living system is also observed in models designed for handling the classical *similia principle*. In a study based on local analogy of symptoms, in this case on the skin, by employing classic models of acute inflammation, the pre-treatment of rats with *Rhus tox* 6CH followed by the induction of inflammatory edema with dextrane – an inductor of mast cell histamine degranulation – did not result in edema inhibition. On the other hand, treatment with a single dose of *Rhus tox* 6CH, at the same moment when dextrane was injected, elicited the inhibition of the edema in levels comparable to pharmacological doses of cyproheptadine. It must be emphasized that in the edema elicited by carrageenan – an agent which mobilizes several chemical inflammatory mediators besides histamine, especially prostaglandins – the results were different: both the pre-treatment and the acute treatment with *Rhus tox* 6CH were able to inhibit the inflammatory edema. An initial evaluation suggests that mast cell degranulation must have had already begun to be receptive to the modulator actions of the ultra-dilutions of *Rhus tox* (Santos et al., 2007).

Memory is another property that seems to be present in the biological processes modulated by ultra-dilutions. It is easily observed in immunological studies. In a model of humoral immune reaction, Weisman et al. (1997) showed that intraperitoneal pre-treatment with diluted antigen (10^{-36}) for one month, followed by immunization of mice with the same antigen in pondered concentrations elicited a humoral immune answer with raise of IgG (memory immunoglobulin); different from the control group, which expressed raise of IgM (primary immunoglobulin). Thus, this study showed that the induction of immunological memory thorough ultra-diluted antigen could be possible.

Another important trait to understand ultra-dilutions is their putative **independence from the pharmacokinetic properties**. Bonamin et al. (2001) showed that the modulator activity of dexametasone 7CH and 15CH on its pharmacological activity is **instantaneous**, i.e. the administration of both preparations (in pharmacological doses and high diluted) in a same syringe elicits simultaneous observable effects. Thus, the presence of ultra-diluted dexamethasone instantly blocked the activity of the same substance in pharmacological doses, as expressed by inflammatory edema and migration of inflammatory cells in models of acute inflammation.

More recently, it has been observed that the **interference by physical factors** (Weber et al., 2007) – especially electromagnetic energy – on ultra-dilutions prepared according the homeopathic pharmacotechnics may determine the disappearance of their modulator effects. The action of microwaves and cell phones on ultra-diluted preparations of thyroxin resulted in the nullification of its inhibiting action on the amphibian metamorphosis, which was easily observed in the control group. This subject is more detailed described in Chapter 10. These findings corroborate earlier studies on the nullifying effects by electromagnetic radiations on ultra-dilutions of histamine (Hadji et al., 1992).

Perhaps, the most enigmatic feature regarding the properties of ultra-dilutions is the **non-linearity** of their effects. In several studies employing *in vivo* and *ex vivo* models, especially involving iso-endopathy, an **oscillatory potency-effect curve** has appeared. The first observations were initially considered as artifacts, but the repetition of this pattern in different studies involving completely different experimental models, in times and places equally different, points out to the existence of a property intrinsic to *dynamized systems*. (Carmine, 1997; Cristea et al., 1997a; Lorenz et al., 2003; Belon et al., 2004; Malarczyk et al., 2003, 2004; Delgado and Ruiz-Vega, 2006).

Among these studies, Cristea et al. (1997) observed the **importance of succussion** in the triggering of this pattern. In this study, successive dilutions without succussion of *Belladonna* resulted in the classic pharmacological effect (dose-effect pattern) up to sixth centesimal dilution, then; no effects were seen at all. On the contrary, dilutions of *Belladonna* with succussions presented a growing oscillatory curve above the threshold of the sixth centesimal dilution. Classic studies show that simple agitation in vortex is able to trigger the specific activity of ultra-dilutions (Davenas et al., 1987).

Non-linearity is even more evident in strict similarity models, but it is also an issue under dispute among authors. In the studies of iso-endopathy models using isolated systems, such as cell and microorganisms (Malarczyk et al., 2003, 2004), the non-linear pattern of reactions is easily described through a potency-effect curve (see Chapter 9 for a more detailed explanation). However, in models of strict similarity, as it happens in the homeopathic clinical practice, the landmark for the observation of the ultra-dilutions effects is the analogy of symptoms, and the biological effects following treatment are in general wide disseminated and expressed in several simultaneous levels.

In human clinical practice, e.g., functional disturbs are observed involving simultaneously several organic systems, associated to behavioral and psychological disturbs, with reflections on social and quality of life aspects (Brunini, 2002). Such disperse effects do not have a common connection thread, a specific molecule-receptor system, as it happens after the administration of a drug. In the latter case, the therapeutic and side effects emerge clearly and simultaneously as a function of the bodily distribution of specific receptors. Contrarily, the changes triggered by a similarity of symptoms are much more general and, thus, apparently unspecific and easily mistaken with a placebo effect, especially in human clinical practice (herein, the possibility of placebo effect of homeopathy is not even considered because of multiple blind studies performed in plants and bacteria that are cited in this book and show clear biological effects that deserve further investigations). The effects of similarity in clinical practice show non-linearity within a notion of **globality** or **totality**. The observation of the general effects of a homeopathic remedy manifested *a posteriori*, associated to the expectancy of some effects *a priori* determined by proving, indicate coherence and convergence among them only when analyzed under an adequate semantic definition of symptoms and proper mathematical models, different from those used in classical pharmacology (Rutten et al., 2003; Rutten, 2007). For details, see Chapter 11.

Traditionally observed in human clinical practice, therapeutics through strict similarity also develops specific changes in the pathological patterns of laboratory animals and plants under several experimental conditions (Oberbaum et al., 1997; Cristea et al., 1997b; Betti et al., 2003; Araújo Prado Neto et al., 2004; Coelho et al., 2006). Similarity, with its properties of non-linearity and globality, the similia principle may be considered a unique biological phenomenon, deserving of more thorough studies.

According to Coelho et al. (2006), the treatment of rats with growing potencies of *Dolichos pruriens* led to a gradual decrease in the itch-behavior elicited by heat, resulting in mild skin lesions due to scratching with the nails. This decrease evolved as a function of time until reaching levels equivalent to the control group, whose animals had not been subjected to heat and, thus, had presented neither itch nor lesions (Fig. 1.1). In this case, the only common aspect between the remedy effect and the itch-behavior was the behavioral analogy. Both the hair that cover the pod of *Dolichos pruriens* and the continual heat elicit itch, however, the physiopathological mechanisms involved in each case are different. In the first instance, the contact with the plant elicits type I hypersensitiveness, with histamine release and stimulation of the local nervous endings. In the second, it is the vasodilatation elicited by continual heat which results in discomfort and grooming, through the action of the central nervous system. Thus, both facts bear no mutual relation from a mechanistic point of view, but only a similarity between the behavioral symptoms. This is the true meaning of the so-called *similia principle*.

This experimental model allows grasping still another fact. The inhibition of the itch-behavior and the consequent preservation of the integrity of the skin did not happen immediately after the administration of the remedy and the animals did not show signs of habituation or down-regulation as it is usually observed during a chronic treatment with anti-histaminic drugs. On the contrary, it was observed in the animals treated with *Dolichos pruriens* a certain "tendency to normality", which evolves slowly and gradually as a function of time and inversely related to the concentration of the supposed active principles which may still be present in growingly smaller amounts. This dynamics would reflect the pattern of activity of the homeopathic remedies in clinical situations, which, as it may be seen, does not correspond to the classic pharmacodynamic pattern.

During the XXth Symposium of GIRI (2006), Madeleine Bastide illustrated in a very clear way the grounds of the *similia principle* (Bastide, 2006). She took an example from virology, establishing a parallel between *vaccinia*, i.e. the systemic reactions triggered by the vaccine prepared from the bovine pox-virus, and the pathological manifestation of *varicela* and *zona*, caused by a herpes virus. In both cases, the skin lesion manifested is a macula-papule that evolves to a vesicle, thus, there is a symptomatic similarity between *vaccinia* and *zona* or *varicela*.

The traditional use of the bovine pox-virus vaccine was to elicit protection through an active immunization of the individuals exposed to human *variola*, since

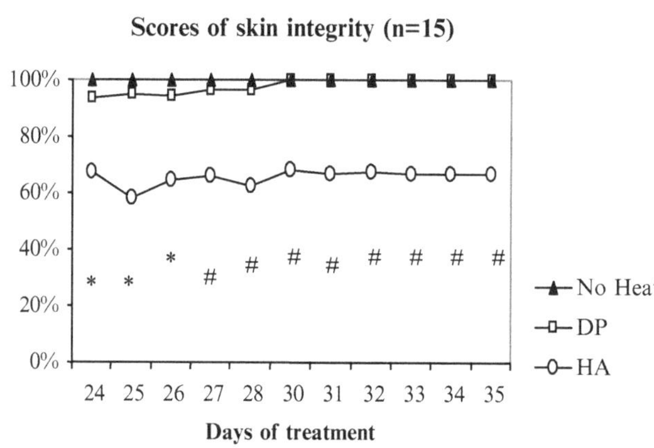

Fig. 1.1 Scores of skin integrity (as percentage) in rats subjected to the induction of itch through continual heat. Each experimental group included 15 animals. The full triangle corresponds to the control animals, not subjected to heat; the square represents the animals treated with *Dolichos pruriens* p.o. in growing potencies (6CH to 12CH); the circle represents the animals subjected to continual heat and treated only with the vehicle employed in homeopathic preparations. (*) represents statistically significant difference ($p \leq 0.05$) between the animals subjected to heat and treated with inert substance and those not subjected to heat. (#) represents statistically significant difference ($p \leq 0.001$) between the animals subjected to heat and treated with inert substance and the other two experimental groups. Tests employed were Kruskal Wallis/Dunn

the virus of bovine and human *variola* present a high antigenic and structural homology. But this vaccine does not protect the same individual against *zona*. On the other hand, the homeopathic remedy *Vaccinotoxinum*, prepared from ultra-dilutions of the bovine pox-virus is traditionally employed as an alternative treatment of *varicela* and *zona*, applying the *similia principle* (Seabra, 1940; Boiron and Payre-Ficot, 2000). Although both are DNA-viruses, the physical and antigenic characteristics of the bovine pox-virus and the herpes virus are completely different, e.g., the former has helicoidally symmetry, while the latter, crystalloid symmetry; they have no antigenic homology at all and there is a large difference between the sizes of the viral particles. There is not the least possibility of one being the result of some genetic mutation from the other; "it would be as if a shark were the mutation of a lion", as said Prof. Bastide. That is to say, the common trait between both DNA-viruses is the morphology of the vesicular lesions.

The **symptom** is the key to the *similia principle*. Although both lesions present mild histological variations, the **body expression** in *variola* and *zona* is so alike as to correspond to the *similia principle*. The most important question remaining is: how is processed the bridge between the changes in the symptoms patterns and the changes in the biological (histo-physiological) patterns that underlie the healing process? There is no immediate answer, and a long path will still have to be threaded before it is reached. In any case, this model allows distinguish the causality in both instances: antigenic homology would not be able to explain the effects of the homeopathic dilutions of *Vaccinotoxinum* in the cure of *zona*, as both viruses have no common elements. This is a simple example of similarity beyond the framework of classic binary causality. Here it is also clear the difference between similarity and isopathy.

To summarize, a global analysis of the main experimental studies involving ultra-dilutions resulted in the identification of certain phenomenological properties and in definitions of some therapeutic models. These are summarized in Table 1.1.

Table 1.1 Observable phenomenological properties and therapeutic models employed in basic research on ultra-dilutions

Therapeutic models	Phenomenological properties
• Isopathy	• Modulator nature
• Iso-endopathy	• Specificity
• Similarity	• Dependence on context and cell organization
	• Dependence on the previous conditions of the sensitive (living) system
	• Memory
	• Instantaneous action
	• Influence of electromagnetic factors
	• Non-linearity
	• Dependence on succession/agitation
	• Totality/globality
	• Tendency to normality
	• Similarity

The Theory of Body Signifiers

The dependence of the sensitive system on the previous conditions, the non-linearity and globality seem to be critical aspects to understand and interpret experimental studies in the field of ultra-dilutions, as they are easily mistaken with artifacts. Thus, it is necessary to conceive lines of thinking that are different from traditional pharmacology, as well as mathematical models able to objectively define the particularities of the results. Concerning the latter aspect, some attempts have been recently made (Zacharias and Zacharias, 1997; Del Giudice and Preparata, 1998; Silva et al., 2005; Betti et al., 2003; Delgado and Ruiz-Vega, 2006; Rutten and Stolper, 2006).

The observation of properties specific to *potentized* systems and the difficulty to design experimental models and to suggest explanatory theories able to give an account of such properties point out to the need of lowering our level of expectations. Thus, for the search of consistent explanations regarding a possible mechanism of action of the ultra-dilutions, it is necessary to start from an even more basic level: the description and understanding of their properties (Bonamin, 2005; Bastide and Lagache, 2007). The aim, then, is to look for a way of thinking the obtained results within a phenomenological approach. These are the grounds on which, along 20 years, the "Theory of Corporeal (*Body*) Signifiers" was built – although the authors prefers call it as an "interpretative hypothesis" (Bastide and Lagache, 1992, 2007; Bastide et al., 1995; Lagache, 1997b).

The acknowledgment of ultra-dilutions as a complex system comes from the concept that some results emerge by the action and interaction of many elements, being, sometimes, unpredictable (Amaral, 2007). Moreover, in this model, the living systems are considered systems operating far from passive equilibrium. A living system is one that continually coordinates actions with the environment, in a recursive manner; thus, it reconstructs itself as long as it is live (Maturana, 1997). This view on the living beings allows for a clear opposition between the autopoiesis of biological regulation processes and purely physico-chemical reactions. As a point of departure, living beings are understood as complex, autopoietic systems, operating in a no linearly network (Le Moigne and Carree, 1997; Bellavite and Signorini, 1998; Bellavite, 2003).

In the 90's, Bastide and Lagache (Bastide and Lagache, 1992; Bastide et al., 1995; Bastide and Lagache, 1997; Lagache, 1997a, b) devoted themselves to the construction of a theory grounded, at the same time, on experimental data and semiotic notions. The result was the "Theory of Corporeal Signifiers", in which the living beings would be understood as complex systems and, moreover, their unique properties might be revealed under the action of the ultra-dilutions. Besides the continual exchange of molecules, a living system would also exchange **non molecular information** with the environment, which would have an effect on its general organization. This idea led to the notion of ***semantic object***, i.e. the representation of biological signifiers. In this way, non-molecular information might modulate biological functions if it is "meaningful" to a particular living system. A ***semantic object*** would allow a fine modulation of biological events through changes in behavioral patterns of different systems, instead of either a direct action or a linear primary modulation – increasing or decreasing the intensity of a phenomenon – as molecules do.

To summarize, according to Bastide and Lagache, the living beings would be not only "open", but also "***sensitive***" systems. Such an approach is directly related to the notion of ***globality*** described above (Bastide, 2001).

The philosophical notion of a semantic object is linked to the hypothesis that there would be a particular path of communication in the living beings, different from the one based on the exchange of objects or molecules, as the mechanistic paradigm states, but that would also be different from the exchange of symbolic information, described as the "paradigm of the sign", at the core of the human sciences. The theory of the corporeal signifiers, thus, proposes a third biological channel of communication, a purely semantic one, instead of strictly biochemical.

The theoretical foundations were built by Agnès Lagache from the questioning about the "original Cartesianism". As it is known, René Descartes (1596–1650) discussed the duality between mind and body and their possible interactions. In the course of time, the original conception of Descartes would have been transformed into two separate, rigid and independent monisms: issues related to the mind, the ideas and the spirit would have had become the subjects for theology and the human sciences, while the issues related to the body and matter in general were the subject of the natural sciences (Lagache, 1997a).

Traditionally, the world of things (matter) is approached by the so-called hard sciences, while the world of ideas is classified as "soft science". Lagache, among other thinkers, suggests a point of departure to review this dichotomy by proposing a common path. Didactically, the theory of the corporeal signifiers would occupy the empty space between these two classic paradigms. Even while bizarre to the eyes of the traditional scientist, this approach fits well in the modern context of **biosemiotics** (Hoffmeyer, 1997; Barbieri, 2007). As remarked by Hoffmeyer "Biology became a kind of no-man's-land between physics and semiotics. Biology must be seen as the interface science, where these two sciences meet".

What Information Is?

According to Bastide and Lagache, the results obtained in the studies on isopathic or homeopathic ultra-dilutions suggest that their modulation on biological events would be processed at different levels of communication (Bastide, 1998, 2001).

At the molecular level, the several cellular interactions may happen at different levels of complexity, which may overlap. This is seen in processes involving drug-receptor linking. The binding of peptides to their specific receptors triggers intracellular chain-reactions able to directly modify its functions and/or activate groups of genes that modulate these functions. This would be considered as a **first level of communication** of the biological systems.

Going further, variations in the complexity of this principle may be observed in physiological conditions. For instance, changes in the concentration of one or more hormones may elicit variations in the direction of their effects. The synergic action of two hormones may trigger an answer in the system completely different from the one triggered by any of them individually, as it is seen in cybernetic regulation (Bonamin, 1994).

Cell-cell and cell-extra cellular matrix interactions also represent equilibrium elements in the cybernetic homeostasis of a particular tissue or organ. A good example to explain it is the process of maturation of T-lymphocytes in the thymus. The selection of immature lymphocytes that pass through the cortical area is effected through the recognizance of the TCR + cells by the epithelial cells. Only the lymphocytes expressing this protein complex pass into the medullar area; the rest is excluded from the parenchyma through apoptosis. In the medullar area, these cells interact with dendritic cells, through the linking between TCR and the protein complex MCH, present on the latter's surface. Cells whose linking is very intense are usually auto-reactive and are also excluded by apoptosis. Several cytokines and thymus hormones participate of the fine adjustment in this stage. Equally, the interactions between proteins in the extra cellular matrix such as laminin, fibronectin and nidogen and the migratory cells also contribute (Savino et al., 2003). To summarize, the maturation of lymphocytes in the thymus is a complex process, involving a multidirectional mosaic of interactions, including different gene expressions, molecules, cells and, especially, different combinatory patterns among them. The final outcome of the process, thus, it is the result of these combinations. Small changes in one or many stages in one or more of the combinations may result in process flaws of cell maturation, leading to immunosupression or autoimmune diseases.

In general terms, it might be said that health and disease represent observable states emerging from combinatory patterns of an immense amount of interacting factors. Susceptibility to disease would, thus, represent the tendency to certain combinations. Hence, the cybernetic regulation of the living systems would constitute a **second level of molecular communication**. Roughly, the behavioral pattern of each element – molecules, genes, cells, matrix – would correspond to one single line in a kind of "bar code" which would define the individuality of the living beings.

Analyzing this standpoint, the following questions arise: what does determine the different combinatory patterns? Are the elements themselves able to draw *a priori* the paths of interaction? Would it be possible to change the combinatory patterns without a necessary biochemical alteration of their elements?

This leads to the notion of **non-molecular information**. For her listeners, to be able to grasp this notion, Madeleine Bastide used to quote the story of Robinson Crusoe as an allegory. When he thought he was lost in a deserted island, Robinson Crusoe found human footprints on the sand, and inferred, "I am not alone". Hence, the footprint is information, the remains of an object which is no longer there but that allowed to establish an important conclusion and a significant change in the state of mind and following actions of the lost hero. Obviously, if instead of a deserted island, Robinson Crusoe had been in Copacabana beach in the summer, he would have never made such an observation. The strength and meaning of a piece of information directly depend on the **context** where it is inserted. Thus, the footprint would be information carrying in itself important meanings, but would not be the **object** or **matrix** itself (i.e., the actual foot of the individual who walked there).

The sand would be the **carrier** of this piece of information. If the sea had risen before Robinson Crusoe arrived at the beach, the information would have vanished. Thus, if the carrier is altered, information may disappear.

Hence, information represents an object but it is not the object itself. In order for information to be perceived and decodified, it is necessary for the receiving system to be sensitive to it, i.e. this representation must be **meaningful** to the receptor (Bastide and Lagache, 2005).

Specificity is, thus, an intrinsic property of the system of information exchange. In the case of homeopathic and isopathic medicine, specificity may be translated into **individuality**. The individualization of the patient is a *sine qua non* factor for therapeutic success, as is exhaustively seen in the clinical practice.

This concept was very well illustrated by Gonçalves et al. (2004). They showed that variations on the degree of individualization might determine different outcomes. Rats with urinary infection were treated with isopathy and similarity-grounded methods. Only the animals treated with isotherapic remedies individually selected and those treated with the remedy prescribed according to the *similia principle* on the grounds of their general behavior (*Phosphorus*) showed decrease of the infection to levels comparable to the antibiotic treatment. This was not observed in the control group or in the animals treated with an isotherapic prepared from pooled urine.

In a recent study (Soares et al., 2007); it was observed that the general activity of rats suffered significant change after a treatment with *Bryonia alba* 200CH. However, the effect was only seen in rats that presented *a priori* decreased general activity, similarly as described materia medica of *Bryonia alba*. In these animals, the general activity increased after *Bryonia alba* 200CH treatment, whose effect was not seen in hyperactive rats, namely, those who did not fit the *Bryonia alba* materia medica descriptions.

These last mentioned studies reveal the parallels between similarity, specificity and individuality. From a practical point of view, this parallel must be an important factor to be taken into account in the formulation of new experimental designs.

From all above, a hypothesis has emerged, suggesting that the universe of ultra-dilutions and particularly of the *similia principle* would operate within an information exchange system. In this sense, the raw material for the preparation of a homeopathic remedy would be the information matrix, the object hosting a set of latent meanings. Water or the solvent employed in the homeopathic preparation would be the information carrier. And the particular living system – or the patient, in the clinical setting – would be the receiver of information. The relevance of the context would be implicit within the notion of similarity. An individual presenting a "Belladonna pattern" would not be able to "understand", for instance "Arsenicum" information. Information would only elicit structural effects on the receiver if meaningful to the latter. Important to explain that it is not a *neural/cognitive* meaning. This is the notion of "semantic object" or the representation of *biological meanings* contained in a substance or a matrix for the homeopathic remedy (Bastide and Lagache, 1997; Lagache, 2005).

In a contextual system, wrong information might potentize already existing disturbs instead of correcting them. In the studies performed by Delbancut et al.

(1993) and Bonamin et al. (2002), undesirable effects of ultra-dilutions would appear as "mistaken" information. In the first case, the pre-treatment of cultured cells with ultra-dilutions of cadmium did not elicited protection, as observed before, but a state of higher cell mortality, if upon the challenge with other substance. In the second, the treatment of rats subjected to chemical hepatocarcinogenesis with ultra-dilutions of dexametasone determined increase in development of preneoplasic nodes and more aggressive phenotype in their constituting cells. On the other hand, no alteration was observed in the rats subjected to simple hepatic regeneration treated with ultra-dilutions of dexametasone (see Chapter 6).

Hence, the proposal by Bastide and Lagache raises biological communication to two further levels: **level three**, corresponding to the principle of identity or isopathy (Delbancut et al., 1997; Datta et al., 2001; Baumgartner et al., 2004); and **level four**, strict similarity (Oberbaum et al., 1997; Araújo Prado Neto et al., 2004; Coelho et al., 2006); both within the context of **non molecular information**. Important to stress, it is a simple proposal, but highly valuable for the researchers who choose to dive into such a difficult field. The adoption of the thinking proposed by Bastide and Lagache has been an amazing tool for the conception of experimental designs and a relative prediction of results. It is not hard to imagine the impact of this theory on the development of the emergent homeopathic science.

As any theory, its validation depends on the corroboration by objective facts and observations. The idea that the water employed as solvent in homeopathic preparations might act as a carrier refers to the need to observe the behavior of perturbations potentially able to "imprint" – through succussion – or "erase" – through electromagnetic fields – the information contained in it. The studies performed by Hadji et al. (1992), Cristea et al. (1997a) and Weber et al., (2007) strongly support this hypothesis (see Chapter 10).

The direction of certain behavioral patterns in the receiver, thus, might be selected by the treatment with ultra-dilutions of compatible substances. However, in more complex settings, the oscillations of the results as a function of the experimental context may be very large. Such oscillations represent the observable state resulting from multiple interactions between a living system and the environment, among which would be included the interferences produced by the ultra-dilutions, as well as other interferences, molecular or not.

For instance, in 1988 Daurat et al. showed that ultra-dilutions of alpha-beta interferon elicited mild immunosuppression in Balb/c mice, but they corrected the humoral and cellular immune answer in NZB mice, which carry lupus. This study clearly shows the importance of the biological system participation in the outcome determination, the dependence on the previous state of the individual in the configuration of the final observable state.

A similar situation was observed when immunosuppressed mice were treated with ultra-diluted complex of thymulin, bursin and IL3: 90% of the treated animals survived, against 10% in the non treated group, when the study was performed in the summer. In the winter, 100% of the treated animals died, while the same 10% in the control group survived. The experiment was repeated 24 times on a monthly basis for two years, and the mathematical treatment of the data showed interference of the circannual rhythm on the action of this isotherapic complex, even when the animals were kept in strict controlled environmental conditions (Guennoun et al., 1997).

This might be the most critical experimental problem: how results with such intense variations may be distinguished from artifact? What are the parameters that must be established in order to be able to safely differentiate the properties of a system of information exchange? A system which comprises important oscillations as a function of the context, how may it be distinguished from mere technical error? For the proposal by Bastide and Lagache to be soundly confronted in the future against new experimental data, it is necessary to clearly define the notion of information and the properties that exclusively emerge from such a system. Table 1.2 shows a schematic representation of the properties of the semantic objects, together with the references of the studies that illustrate them. It is hoped that this table might serve to stimulate the construction of experimental models designed in order to reveal all the nuances and peculiarities intrinsic to the ultra-dilutions.

In Short ...

The notions of autopoiesis and structural closure of the living systems proposed in the '60s allow understanding some aspects of their complexity. To Maturana, structural changes in the organism are not determined by the environment, but by its own structure (Maturana, 1997). For instance, the organic reactions following a bacterial infection would not be *determined* but *triggered* by the bacteria. What would determine

Table 1.2 Phenomenological properties of "semantic objects", their correspondence to the theory of corporeal signifiers and experimental illustrations

Phenomenological properties of the "semantic objects"	Experimental illustrations
• Modulator nature	• Sainte-Laudy and Belon, 2006
• Specificity	• Gonçalves et al., 2004
• Dependence on context and cell organization	• Youbicier-Simo et al., 1996, 1997; Bonamin et al., 2002
• Dependence on the previous conditions of the sensitive (living) system	• Walchli et al., 2006
• Memory	• Weisman et al., 1997
• Instant action	• Bonamin et al., 2001
• Influence of electromagnetic factors	• Hadji et al., 1992; Weber et al., 2007
• Non-linearity	• Delgado and Ruiz-Vega, 2006; Malarczyk et al., 2003, 2004
• Dependence on succussion/agitation	• Cristea et al., 1997
• Totality/globality	• Rutten, 2007
• Tendency to normality	• Coelho et al., 2006
• Similarity of symptoms	• Oberbaum et al., 1997; Araújo Prado Neto et al., 2004

such reactions – fever, pain, weariness, etc. – could be the structure of the particular body which suffers the infection. Indeed, a same bacterial toxin, in a certain concentration, may trigger a temporary fever in a human individual, be innocuous to a fish or immediately lethal to some species of crabs (*Limulus polyphemus*) (Berczi, 1998). Such differences, thus, would be subordinated to the different patterns of interaction that the inflammation chemical mediators and other elements develop in each species or individual. A microorganism, in this case, would merely "press the button". Curiously, such an observation may be compared to the notion of susceptibility described by James T. Kent in the beginnings of the 20th century: "Do not say that a patient is ill because he has a swelling, but rather, that the swelling is there because the patient is ill" (Kent, 2002).

The notions proposed by Bastide and Lagache emerge from this conception and go beyond it. Could high dilutions model the output pattern of a being after some stimuli have *pressed its buttons*? Agnès Lagache describes this dynamics as it follows:

> *Living beings are not things: concepts made for things and from things cannot be appropriate for living behavior. We build our own notions....*

Note on Concepts Mentioned in This Chapter

Agnès Lagache, January 2008.

From many fronts, today, new ways of dealing with living beings arise: a new vocabulary is coming in use, too often without really defining the implied notions, and this can lead to misunderstandings.

First of all, the notion of **totality** becomes a kind of common place. As a matter of fact, living beings in their capacity behave as totalities; it is a mere evidence of observation; but reasoning in terms of totality is opposed to the uses of analytic classical science; whence the difficulty...

Another example is the vocabulary analogically borrowed from the area of **information** and **communication**: imported and made usual from genetics, where it has a very specific meaning, it becomes trite in many biological fields as much as living beings obviously deal and exchange incessantly with their environment. But it is necessary to define acutely what is called information and communication: as it goes nowadays, if everything is so-called information, then nothing would be information; one must especially distinguish carefully whether it is exchange of objects or exchange of pieces of information, because the rules of exchange and their effects are totally different in each case.

Information and totality lead to **specificity**: it is no more than a logical inference, but must be explained.

As we enter the new field of communication, we must also understand that different kinds and **levels** of informative phenomena can take place; confusing them leads to blind interpretation of experiments.

So I will try to answer four questions, or better say to open some track in the way of answering them:

1. What do we call **totality**? In what sense a living being appears unified?
2. What is an **informative** event? And what is it not?
3. Why each situation for a living being is a **specific** one?
4. What are the main different **levels** of informative facts that we can already sketch?
5. I will conclude by specifying the status and the limits of these conceptual tools.

1. **What do we call totality?**

We call totality the fact that a living being acts and reacts with its whole dynamic organization, whose elements are strongly linked; this act makes sense confronting an environmental situation. The unity of the living being stands in its dynamic capacity which makes each part work with every other and globally answer the situation. In this view, each local reaction of part of the organism must be seen as a beacon for the whole working, and each local fact is linked with others and must be considered amongst them in a global grasp.

This totality makes sense for an external observer: it is behavior. Any action commits the whole system to global behavior: deal, or symptom, it is understandable as a meaningful behavior confronting environment.

This fact can be also described as the result of strong binding between external and internal events for living beings: if we describe things as resulting from a double set of internal interactions and external ones, we can see a difference between inanimate objects and living ones. A stone is as it is because of its internal molecular structure; it exchanges with environment, for example caloric exchanges and pressure, and the totality of these exchanges determine what the stone is at the moment. But in the limits of ordinary circumstances, the influence of external facts is very weak and does not determine the state of things, which are much more determined by their internal composition. So we would not consider them as totalities because they can be understood by the mechanical set of their elements.

On the contrary, a living being is much more determined by its external exchanges with environment, which have immediate consequences in the internal world: to adapt, to answer, is to modify the self with regard to exchange. Hence the internal world, incessantly renewed and changed according to exchanges, is much more determined by these exchanges than by internal elements, in the limits of its structure.

So, to summarize: each entity results from its internal and external interactions; for inanimate things, internal interactions are generally more important than external ones; for living beings, external interactions are more heavily determinant than internal ones, in the scope of its structure.

So the idea of totality arises as an essential feature of living beings, because not only internal elements work together but this internal set is also linked with external links, and the living individual is in some way the idiosyncratic "knot" of these whole exchanges.

2. **What is an informative event? And what is it not?**

The main confusion about information comes from the fact that we reason about information as we reason about things. So the main point is that information is not a thing; it can be anything, a change in a structure, a position, a form, anything which becomes meaningful according to the ability of a reader.

In the area of language and consciousness, information are more often in words, but also in gestures, voice, doing. It requires a reading mind to become meaningful.

In biological area, nothing of this kind, but an ability of the organism to use some elements as regulation signals, at different levels of informative relationship.

So we have two rules there: (1) A piece of information is not an object; it can be a simple change in a previous structure. Exchanging objects, the receiver does not determine the received object; exchanging pieces of information, the receiver determines what information is for him according to its own circumstances. (2) Never do a mix-up between the carrier of a piece of information and the proper informative effect, the "message", which exists only if is read by the receiver.

3. **Why each situation for a living being is a specific one?**

Each state of living being is therefore a specific one: it is a synthesis, the result of all historical events which molded him. So each new situation is also for him a global face of the world demanding to deal with; it cannot be passive, it cannot be deaf. Whether in the form of a successful adaptation or by a symptom, there will be an answer. This answer is also one global answer, because the whole being is engaged.

So if you want to modify this answering, it must fit exactly with this synthetic state of the organism: information does not bear approximation; this piece of information must be the exact correlative of the whole symptoms in order to induce this resonance with a dynamic effect.

To summarize: each state of a living being is unified as it is a synthesis; so it must be dealt with as a whole, and this whole is original as it results from the genuine story of each organism.

Fortunately, there is some redundancy in the behavior of living beings, because environmental circumstances are not in infinite number, and possibilities to answer are also limited taking into account the common abilities of living beings. It makes possible to have a typology of recognizable behaviors; but the list is never totally closed...and so widely open that it is not easy to recognize the proper remedy for the proper subject. As Professor Leoni Bonamin beautifully says, *for animals, specificity equals individuality*.

4. **What are the main different levels of informative facts that we can already sketch?**

This point has not been anticipated by theoretical approach but has little by little emerged from the acute observation of experimental results by Professor Madeleine Bastide.

Everything happens as if we might distinguish between different levels of informative facts, according to the kind of information used. I will comment the four levels settled by Professor Leoni Bonamin:

- At the first level (molecular level, cybernetic regulation), I wonder if it is really relevant to speak of communication: cybernetic regulation uses some threshold as a **signal,** inducing feed-back. But the chain of events is completely molecular in the classical way of interacting.
- At the second level, we encompass much wider interactions, but it may be a large combination of simple facts, according to a rather molecular model of key/keyhole interaction.
- The third level enters information area; there is no more a causal object in the classical way, that is: information can be a high dilution. So it is experimentally ambiguous: the presentation as a high dilution might be enough to give to the substance an informative effect, or not enough to differentiate it from the molecular product already known by the organism. The sensitivity to context is increased. For example, in clinical use of *Thymulin*, physicians report frequent aggravation instead of healing; but if they "surround" *Thymulin* by an individualized remedy, the effect is O.K. It fits with an informative interpretation: the same does not inform the same.
- It is at the fourth level that we are really at ease with informative interpretation. The remedy has two good features of a piece of information. Everything happens as if dilution gave it the form of an image of object, at least a kind of representative of the object; and moreover it has an exogenous origin, which prevents organism to confound it with self. So it happens that it acts as a piece of information about the proper state of the organism, a critical image of its state, and induces the organism to deal afresh with this state. The dynamical effect depends both on the similarity which induces a resonance with the proper state of the organism and of the dilution which provides not the object but some representative of it.

5. **Epistemological status and the limits of these conceptual tools.**

Here are the main features of this paradigm of corporal signifiers that we used with Madeleine Bastide during the past 20 years.

Is it a theory? I don't think so; I would rather prefer to call it hypothesis, because it is not done to speak about the nature of facts but to interpret and, to some extend, predict experimental data. It does not explain anything, in the sense that it does not pretend to say anything about the *things (*organism, carrier, informative structure); it just says something about their mutual behavior.

In a word, it is an interpretative hypothesis and not an explicative theory.

In this way, it is totally neuter facing materialism or vitalism. It just assumes that biological material organizations are able to deal with information, although we believed it was only the fact for mind. If you choose a spiritual way of thinking, you may consider that something of the biological device is already close to consciousness, as Bergson did. If you prefer the materialistic way of thinking, as I do, you will, on the contrary, note that mind is a biological device, and that's not a scoop...

In any case, it does not concern the use of the hypothesis in order to understand better experiment. This use is, up to my opinion, a temporary and operating way of thinking, in order to experiment and observe; I think it will be overtaken when we are able to build a sound theory of life, but may be it will have contributed a little to build it. And that is the task now: to experiment according to the rules imposed by the objects as the biologist deals with.

Acknowledgements Thanks to Dr. Maria Silvia Furquim de Azevedo Morgulis for reviewing the text and Dr. Silvia W. Priven for comments and translation.

References

Amaral MTCG. Uma visão de ciência. Lecture at UNISA. May, 31st, 2007; São Paulo, Brazil.

Araújo Prado Neto J, Perazzo FF, Cardoso LGV, Bonamin LV, Tavares Carvalho JC. Action of Causticum in inflammatory models. Homeopathy; 2004, 93: 12–16.

Atlan H. L'argument d'éfficacité: médecine scientifique contre nostalgie scientifico-mystique. Annales Les Entretiens Internationaux de Monaco – validation scientifique des medecines non conventionnelles – L'Example de l'homéopathie. Mars, 2007; Monte Carlo, Monaco.

Barbieri M. (Ed). Introduction to biosemiotics – The new biological synthesis. Dordrecht: Springer; 2007. ISBN-10 1-4020-4813-0 (HB).

Bastide M. Information and communication in living organisms. In: Schulte J, Endler PC (Eds). Fundamental Research in Ultra High Dilution and Homeopathy. Dordrecht: Kluwer; 1998.

Bastide M. Globalité, Homeopathie et communication dans le vivant. Aesculape; 2001, 28: 6–10.

Bastide M. Interpretative theories about ultra-dilutions: Supporting evidences. Annals of XX Symposium of GIRI. Cultura Homeopática; 2006, 16: 22–30.

Bastide M, Lagache A. The paradigm of signifiers. Paris: Atelier Alpha-Bleue; 1992.

Bastide M, Lagache A. Communication process: A new paradigm applied to high-dilution effects on the living body. Alther. Ther. Health Med.; 1997, 3(4): 35–39.

Bastide M, Lagache A. The information theory by corporeal signifiers applied to experimental research in unconventional medicines. In: Halm R (Ed). Unconventional Medicine in Front of Science – a New Approach to Health – Body Intelligence. Monte Carlo, Monaco: Les Entretiens Internationaux de Monaco Editions; 2005. ISBN. 2-915664-01-9.

Bastide M, Lagache A. Analyse critique des hypothèses explicatives de l'homéopathie. Annales Les Entretiens Internationaux de Monaco – validation scientifique des medecines non conventionnelles – L'Example de l'homéopathie. Mars, 2007; Monte Carlo, Monaco.

Bastide M, Daurat V, Doucet-Jabeuf M, Pelegrin A, Dorfman P. Immunomodulator activity of very low doses of thymulin in mice. Int. J. Immunother.; 1987, 3(3): 191–200.

Bastide M, Lagache A, Lemaire-Misonne C. Le paradigme des signifiants: Schème d'information applicable en immunologie et en homéopathie. Révue Internationale de Systémique; 1995, 9(2): 237–249.

Baumgartner S, Thurneysen A, Heusser P. Growth stimulation of dwarf peas (*Pisum sativum* L.) through homeopathic potencies of plant growth substances. Forsch Komplementarmed Naturheilkd.; 2004, 11: 281–292.

Bellavite P. Complexity science and homeopathy: A synthetic overview. Homeopathy; 2003, 92: 203–212.

Bellavite P, Signorini A. Pathology, complex systems and resonance. In: Schulte J, Endler PC (Eds). Fundamental Research in Ultra High Dilution and Homeopathy. Dordrecht: Kluwer; 1998.

Belon P, Cumps J, Ennis M, Mannaioni PF, Roberfroid M, Sainte-Laudy J, Wiegant FAC. Histamine dilutions modulate basophil activation. Inflamm. Res.; 2004, 53: 181–188.
Berczi I. Neurohormonal host defense in endotoxin shock. In: Neuroimmunomodulation – molecular aspects, interpretative systems and clinical advances. Ann. N. Y. Acad. Sci.; 1998, 840: 787–801.
Betti L, Lazzarato L, Trebbi G, Brizzi M, Calzoni GL, Borghini F, Nani D. Effects of homeopathic arsenic on tobacco plant resistance to tobacco mosaic virus. Theoretical suggestions about system variability, based on a large experimental data set. Homeopathy; 2003, 92: 195–202.
Boiron M, Payre-Ficot A. A homeopatia no dia a dia. Lyon: Ed. Boiron; 2000.
Bonamin, L.V. O estresse e as doenças. Ciência Hoje: 1994, 17(99): 24–30.
Bonamin LV. Experimental Illustrations of body signifiers theory. In: Halm R (Ed). Unconventional Medicine in Front of Science – a New Approach to Health – Body Intelligence. Monte Carlo, Monaco: Les Entretiens Internationaux de Monaco Editions; 2005. ISBN. 2-915664-01-9.
Bonamin LV, Abel MNC. Histamina, Serotonina e seus antagonistas. In: Helenice de Souza Spinoza, Silvana Lima Gorniak, Maria Marta Bernardi (Eds). Farmacologia Aplicada à Veterinária. 3° ed. Rio de Janeiro: Guanabara Koogan; 2002.
Bonamin LV, Martinho KS, Nina AL, Caviglia F, Do Rio RGW. Very high dilutions of dexamethasone inhibit their pharmacological effect *in vivo*. Br. Hom. J.; 2001, 90: 198–203.
Bonamin LV, Martinho KS, Marcondes VA, Benvenga GU. Ultra-high dilutions and proliferative activity of liver. Annals of the GIRI meeting. October, 2002; Monte Carlo, Monaco.
Brunini CRD. Qualidade de Vida e Abordagem Homeopática em Crianças Asmáticas. Infanto – Neuropsiquiatria da Infância e Adolescência; 2002, 10(1): 18–21.
Carmine TC. Effects of high potencies of tumor necrosis factor alpha on H_2O_2 production in cultured neuroblastoma cells by enhanced luminol-dependent chemiluminescence (ECL). Br. Hom. J.; 1997, 86: 67–72.
Coelho CP, D'Almeida V, Pedrazzolli Neto M, Duran-Filho C, Florio JC, Zincaglia LMC, Bonamin LV. Therapeutic and pathogenetic animal models for *Dolichos pruriens*. Homeopathy; 2006, 95: 136–143.
Cristea A, Nicula S, Darie V. Pharmacodynamic effects of very high dilutions of *Belladonna* on the isolated rat duodenum. In: Bastide M (Ed). Signals and Images. Dordrecht: Kluwer; 1997 (a).
Cristea A, Teodorescu-Negres S, Darie V. *Chamomilla* homeopathic diluition effect on central nervous system. In: Bastide M (Ed). Signals and Images. Dordrecht: Kluwer; 1997 (b).
Datta SS, Mallick PP, Khuda-Bukhsh AARR. Comparative efficacy of two microdoses of a potentized homoeopathic drug, cadmium sulphoricum, in reducing genotoxic effects produced by cadmium chloride in mice: A time course study. BMC Complementary and Alternative Medicine; 2001, DOI 1472-6882/1/9.
Daurat V, Dorfman P, Bastide M. Immunomodulatory activity of low doses of interferon α,β in mice. Biomed. & Pharmacother; 1988, 42: 197–206.
Davenas E, Poitevin B, Benveniste J. Effect on mouse peritoneal macrophages of orally administered very high dilutions of silica. Eur. J. of Pharmacol.; 1987, 135: 313–319.
Davenas E, Beauvais F, Amara J, Oberbaum M, Robinzon B, Miadonna A, Tedeschi A, Pomeranz B, Fortner P, Belon P, Sainte-Laudy J, Poitevin B, Benveniste J. Human basophil degranulation triggered by very dilute antiserum against IgE. Nature; 1988, 333: 816–818.
Delbancut A, Merlet D, Mellado M, Dorfman P, Cambar J. Ètude de L'effet Protecteur de Très Hautes Dilutions de Cadmium vis-à-vis de Concentrations Cytotoxiques de ce Même Métal sur des Cultures Cellulaires Rénales. Cahiers de Biothérapie; 1993, 120: 39–44.
Delbancut A, Barouillet MP, Cambar J. Evidence and mechanistic approach of the protective effects of heavy metal high dilutions in rodents and renal cell cultures. In: Bastide M (Ed). Signals and Images. Dordrecht: Kluwer; 1997.
Delgado GE, Ruiz-Vega G. Non-linear modelling of ultra-dilutions. Annals of XX Symposium of GIRI. Cultura Homeopática; 2006, 16: 31–34.
Del Giudice E, Preparata G. Coherent electrodynamics in water. In: Schulte J, Endler PC (Eds). Fundamental Research in Ultra High Dilution and Homeopathy. Dordrecht: Kluwer; 1998.

Endler PC, Pongratz W, Smith CW, Schulte J. Non-molecular information transfer form thyroxin to frogs with regard to homeopathic toxicology. Vet. Hum. Toxicol.; 1995, 37(3): 259–260.
Gonçalves MI, Biagine M, Neto MF, Schor N. O uso da homeopatia no tratamento da infecção urinária em ratas. Anais do VIII SINAPIH. May, 2004. São Paulo, Brazil.
Guedes JRP, Ferreira CM, Guimarães HMB, Saldiva PHN, Capelozzi VL. Homeopathically prepared dilution of *Rana catesbeiana* thyroid glands modifies its rate of metamorphosis. Homeopathy; 2004, 93: 132–137.
Guennoun M, Boudard F, Cabaner C, Robbe Y, Dubois JB, Bastide M. Radioprotection and immune system regeneration of irradiated mice by using high dilution treatment. Chronobiol. Intern.; 1997, 14(Suppl.1): 60.
Hadji L, Arnoux B, Benveniste J. Effect of dilute histamine on coronary flow of guinea-pig isolated heart. Inhibition by a magnetic field. Abstract. FASEB J; 1992, 5:A1610.
Hahnemann S. Organon da Arte de Curar. 6ª Ed. São Paulo: Robe Ed.; 1996.
Hirst SJ, Hayes NA, Burridge J, Pearce FL, Foreman JC. Human basophil degranulation is not triggered by very dilute antiserum against Human IgE. Nature; 1994, 366: 525–527.
Hoffmeyer J. Biosemiotics: Towards a new synthesis in biology. Eur. J. Semiotic Studies; 1997, 9(2): 355–376.
Kent JT. Lições de Filosofia Homeopática. 2ª Ed. São Paulo: Ed. Organon; 2002. Tradução de Célia Regina Barollo.
Lagache A. Notes on conceptual basis of science. In: Bastide M (Ed). Signals and Images. Dordrecht: Kluwer; 1997 (a).
Lagache A. What is information? In: Bastide M (Ed). Signals and Images. Dordrecht: Kluwer; 1997 (b).
Lagache A. Wordless body – but meaningful – searches metaphor, extra options if mutual linking. In: Halm R (Ed). Unconventional Medicine in Front of Science – a New Approach to Health – Body Intelligence. Monte Carlo: Les Entretiens Internationaux de Monaco Editions; 2005. ISBN. 2-915664-01-9.
Le Moigne JL, Carree H. On modelling complex system applied to homeopathy. In: Bastide M (Ed). Signals and Images. Dordrecht: Kluwer; 1997.
Lorenz I, Scheider EM, Stolz P, Brack A, Strube J. Sensitive flow cytometric method to test basophil activation influenced by a homeopathic histamine dilutions. Forsch Komplementärmed Klass Naturheilkd; 2003, 10: 316–324.
Malarczyk E, Jarosz-Wilkołazka A, Kochmańska-Rdest J. Effects of low doses of guaiacol and ethanol on enzymatic activity of fungal cultures. Nonlinearity in Biology Toxicol. and Medicine; 2003, 1(2): 176–184.
Malarczyk E, Kochmanska-Rdest J, Pazdzioch-Czochra M. Effect of low and very low doses of simple phenolics on plant peroxidase activity. Nonlinearity in Biology, Toxicol. and Medicine; 2004, 2: 129–141.
Maturana HR. Everything is said by an observer. In: Thompson WI (Ed). Gaia: A Way of Knowing. Political Implications of the New Biology. New York: Lindisfarne Press; 1997. Tradução de Nelson Vaz.
Oberbaum M, Weisman Z, Kalinkovich A, Bentwich Z. Healing chronic wounds performed on mouse ears using silica (SiO_2) as a homeopathic remedy. In: Bastide M (Ed). Signals and Images. Dordrecht: Kluwer; 1997.
Priven SW. Além do princípio da vida. Cultura Homeopática; 2005, 12: 15–20.
Rutten ALB. Bayesian homeopathy: Talking normal again. Homeopathy; 2007, 96: 120–124.
Rutten ALB, Stolper CF, Lugten RFG, Barthels RWJM. Assessing likelihood ratio of clinical symptoms: Handling vagueness. Homeopathy; 2003, 92: 182–186.
Rutten L, Stolper E. "Proof" against homeopathy in fact supports homeopathy. Homeopathy; 2006, 95: 57–61.
Sainte Laudy J, Belon P. Improvement of flow cytometric analysis of basophil activation inhibition by high histamine dilutions. A novel basophil specific marker CD 203c. Homeopathy; 2006, 95 (1): 3–8.

Santos AL, Perazzo FF, Cardoso LGV, Carvalho JCT. *In vivo* study of the anti-inflammatory effect of *Rhus toxicodendron*. Homeopathy; 2007, 96: 95–101.
Savino W, Ayres-Martins S, Neves dos Santos S, Smaniotto S, Ocampo JSP, Mendes da Cruz DA, Terra Granado E, Kusmenok O, Villa Verde DMS. Thymocyte migration: Na affair of multiple cellular interactions? Braz. J. Med. Biol. Res.; 2003, 26: 1015–1025.
Seabra A. Medicina popular: sarampo, catapora. Revista de Homaopatia; 1940, 5: 309–11.
Silva WRG, Rodrigues SCP, Storace JEO, Carvalho JCT. Ultra high dilutions and its virtual structures. Arztezeitschrift fur Naturheiverfaren; 2005, 46(10): 614–618.
Soares FM. Atividade geral de ratos tratados com *Bryonia alba* 200CH. Trabalho de Conclusão de curso de graduação: Universidade de Santo Amaro, Brasil; 2007.
Walchli C, Baumgartner S, Bastide M. Effect of low doses and high homeopathic potencies in normal and cancerous human lymphocytes: an *in vitro* isopathic study. J. Altern. Complement. Med.; 2006, 12(5): 421–427.
Weber S, Welles SU, Suanjak-Traidl E, Scherer-Pongratz W, Frass M, Endler PC, Lothaller H. Influence of electromagnetic fields on the effect of homeopathically prepared thyroxin (10^{-30} parts by weight) on highland frogs. Annals of the XXI Symposium of GIRI. March, 2007. Monte Carlo, Monaco.
Weisman Z, Oberbaum M, Topper R, Harpaz N, Bentwich Z. High dilutions of antigens modulate immune response to KLH. In: Bastide M (Ed). Signals and Images. Dordrecht: Kluwer; 1997.
Wiegant FAC. Memory of water revisited. Nature; 1994, 370: 322.
Youbicier-Simo BJ, Boudard F, Mékaouche M, Baylé JD, Bastide M. A role for Bursa Fabricii and bursin in the ontogeny of pineal biosynthetic activity in the chicken. J. Pineal Res.; 1996, 21: 35–43.
Youbicier-Simo BJ, Boudard F, Guellati M, Mekaouche M, Baylé JD, Bastide M. The role of the Bursa de Fabricius and highly dilute bursin in immunoendocrine interactions in the chickens. In: Bastide M (Ed). Signal and Images. Dordrecht: Kluwer; 1997.
Zacharias CR, Zacharias AC. Physical modelling of dynamization. Br. Hom. J.; 1997, 86: 207–210.

Part II
Basic Research: Physics

Chapter 2
Theoretical and Physical-Chemical Models for Dynamized Systems: Validation Criteria

Carlos Renato Zacharias

Introduction

Whenever one thinks about how Homeopathy works, one of the first thoughts is the water memory hypothesis. So, why this idea is so common, what does water memory means and how important is this hypothesis? These questions are relevant to understand the science and culture behind the homeopathic phenomenon.

In order to better understand Homeopathy, one can separate two complementary aspects: clinical and fundamental. The first was systematized by Hahnemann in the 18th century and the clinical concepts and practices have been applied and refined since 200 years ago. The theoretical support used on clinics can be found in classical books like the *Organon* and many versions of the *Materia Medica* as well as in clinical evidence, obtained in agreement with rigorous methodology, can be found elsewhere (Dantas and Rampes, 2000; Singer et al., 2007; Oberbaum et al., 2003). So the clinical aspect is well established and permits to spread the homeopathic therapeutics and culture, all over the world.

However, the scientific aspect is not yet supported by the current established formal science. To analyze this point, one first has to take a view on homeopathic medicine as *potentized* system medicine. Any substance, organism or system can be prepared through a serial process of dilution and agitation (named as dynamization). Thus, one obtains what can be called a *potentized* system the main characteristics of which are the agitation history and a high dilution degree, generally crossing Avogadro's limit. In order to classify a *potentized* system as a homeopathic medicine it one step more is required: a pathogenetic trial. After record of the symptoms, in a well controlled trial (Dantas et al., 2007) it can lead us, accordingly the principle of similarity, as a homeopathic medicine. Semiotics approaches to medicine-patient-physician relationship can be found elsewhere (Jurj, 2007) and is out the scope of this article.

Universidade Estadual Paulista, Guaratinguetá, Brazil. zacha@feg.unesp.br
Presented at the XXth Symposium of GIRI, São Paulo, 2006

L.V. Bonamin (ed.) *Signals and Images*,

Such distinctness is important to conceive, perform and interpret physical-chemical experiments on Homeopathy. As far the author knows, there is not a single experiment that correlates physical-chemical variations with biological response. Also, such experiments are never conceived using the principle of similarity, because they involve machines only, where symptoms are a nonsense property. Thus, what kind of information one can obtain from a physical-chemical experiment and how it can be related to Homeopathy? This article discusses such points aiming to put some light over on the topic, suggesting optimized approaches to study the science behind homeopathy, as well as validation criteria for physical-chemical models.

Methods

In order to discuss the relevance of physical-chemical experiments and the water memory hypothesis to explain homeopathic phenomena, one has to answer some basic questions: What does water memory mean? Why is this idea so common? How important is this hypothesis? What kind of information can one obtain from a physical-chemical experiment? How to relate such experiments to Homeopathy? What kind of experiment is more adequate to study a *potentized* system? These questions are important to those interested in the scientific aspect of Homeopathy.

Results

Water memory represents a controversial property associated to water (or aqueous systems); able to retain in its dynamics and/or structure some correlation with the way it was prepared. Such memory has been observed comparing physical-chemical and biological properties of normal against *potentized* systems. The differences can be evidenced by different biological responses or physical-chemical changes, associated to cluster dynamics or some collective effect.

Considering that the preparation of a *potentized* system involves a process of serial dilution that leads to concentrations above Avogadro's limit, where molecules present in the initial solution can not be found in the final solution, the focus lies on the phenomena linked to the solvent.

The experimental evidences about the phenomenon have been collected by independent groups, using different techniques. Recently, two experimental groups have produced many results, suggesting some theoretical explanations.

Elia and collaborators (Elia et al., 2005, 2006, 2007) have been studying intensively calorimetric and electrical properties. These experiments revealed slow water dynamics dependent on the way the samples are prepared as well as on the substance initially present. They observe a time dependent increment on the "heat of mixing" (Q_{mix}) and an excess of specific conductivity (χ^E), as well a linear correlation between both parameters. The interpretation of these results suggests that the dynamization procedure leaves the aqueous system in a far from equilibrium state, where the presence of dissipative structures would define the main behavior of the system. Theoretical approaches behind this rationale are derived from Prigogine's work (Prigogine, 1978), hopping mechanism and thermodynamics of irreversible processes. All these theories really deal with dynamical effects, temporal evolution and non-minimum energy states, but do not obviously seem to be related to the homeopathic phenomenon.

Rey (Rey, 2003, 2007), van Wijk and collaborators (Van Wijk et al., 2006) have shown differences in thermoluminescence of high dilutions when compared with the respective diluent alone. The working hypothesis is that hydrogen bonds are affected by dynamization. To test this, the authors investigated the low temperature thermoluminescence of high dilutions prepared from LiCl and NaCl solutions, using heavy water (deuterium oxide – D_2O). Such technique is very sensitive in detecting trace substances and would indicate structural changes induced by dynamization, related to the starting substance. The observed temperature peaks may be related to structural phase transition, but the details seem to reveal some memory effect. These results were extensive and more studies have to be done, before a stronger conclusion can be established. Even if the current results were conclusive, any correlation would be linked to the homeopathic phenomenon.

Theoretical models also have been proposed, generally based on quantum concepts. The absence of the starting molecules (mother tincture) on diluted solutions brings one to think of non local effects and entangled states, in order to explain biological effects (Milgrom, 2007; Weingärtner, 2007). But these models require the weakening of quantum theory (Atmanspacher et al., 2002) in order to expand it to non-physical systems and no experimental verification can yet be proposed.

In some models, electromagnetic fields seem to be important to support the biological evidences (Endler et al., 1994, 1995, 1997). In others, clathrates formations are conceived as information host (Anagnostatos, 1994). Nanobubbles (Rey, 2007) and silica chains (Anick and Ives, 2007) have also been suggested to explain theoretical hypotheses and experimental evidences. But these hypotheses cannot explain why and how *potentized* systems produce biological responses. Similar ideas, models and methodologies can be found in some other works (Aabel et al., 2001; Demangeat et al., 1992; Lobyshev et al., 1999; Samal and Geckeler, 2001; Klimek et al., 2005). Despite the proposition of many experimental and theoretical models, very little advance has been done toward the comprehension of the *potentized* systems action and of the homeopathic phenomenon.

Would this crisis be a symptom of some conceptual mistake?

Discussion

Whenever one desires to measure physical properties of a *potentized* system, to compare them with a control sample, the experimental set up is adjusted to highest sensibility. This is a logical procedure because one is expecting small differences.

In these situations, the signal-noise ratio decreases a lot, and the methodological rigor becomes very important. The measurement of very weak signals is really difficult and intense statistics have to be performed to validate the results. But sensibility is not specificity!

The common procedure to measure the mass of an object is using a balance. But no information can be extracted e. g. about the color, because even a high accurate analytic balance is not able to measure electromagnetic fields. The crucial point in this crude example is the equipment specificity, not the sensibility. What does it mean, or what are the consequences of equipment specificity?

First one has to assume that we are looking for a phenomenon not yet well described by formal science. Hypothetically, let's name the homeopathic active element as "information" (Bastide et al., 1995; Lagache, 1997a, b). If the nature of such "information" or its carrier is not electromagnetic, no spectroscopic experiment can measure it directly. So, no evidence could be realized using NMR, luminescence, Raman, UV, IR, and similar techniques. Up to now, there is not even consensual evidence about the electromagnetic nature of this "information", as well its carrier. If the information could be described by its effects on solvent structure, the current experimental spectroscopic resolution would be high enough to detect them. The same reasoning can be expanded to electrical and calorimetric studies, as well as any other physical-chemical investigation. Further, there is the problem of biological specificity, demonstrated on proving trials and the homeopathic clinics: such specificity might be memorized on the solvent, in many different patterns, which has never been observed.

But, if we are unable to detect such information or its effects using physical apparatus, how can one explain the currently published results?

Pure water dynamics is really intense and surprising. When submitted to dynamization, new forces come into action, producing unusual effects. But pure water is not a reality on homeopathic pharmacies. In fact, nor in a research laboratory, because water contains a number of molecular species, ionic contaminants, dissolved gasses and solutes, aerosols, silica, among others substances (Teixeira, 2007; Chaplin, 2007). Thus, experimental conclusions must consider such natural dynamics and intrinsic species, other than contaminants. The higher the sensitivity of an experiment, the more intense is the experimental artifacts.

The water cluster hypothesis also can be put to discussion. Water structure can be thought statistically, as an averaged density fluctuation (Teixeira, 2007), because hydrogen bonds last for a fraction of seconds, instead of an averaged structure that can be recognized like in solid water (ice) and metal hydrates (Chaplin, 2007). Further, the presence of ethanol molecules increases the system complexity.

So, there are so many ways to interpret the experimental results about aqueous solutions, based on the formally established science knowledge (Chaplin, 2007), that one must take care about unnecessary hypotheses or fragile conclusions.

Further, current physical-chemical evidences fail in one critical aspect: there are not enough experiments correlating such evidences with biological responses.

The homeopathic phenomenon seems to be observed only in the presence of a biological sensor. Clinical evidences, proving trials, experiments on embryos, plants, animals, bacteria, and other living systems are able to show the homeopathic phenomenon (Bastide et al., 1985, 1987; Bonamin et al., 2001; Brizzi et al., 2000; Youbicier-Simo et al., 1993, 1996a, b; Betti et al., 1997). Of course, it is not easy to perform or reproduce such experiments due to the intrinsic complexity of life systems as well as the principle of similarity considerations. But, all conclusive results about the phenomenon have been extracted from biological assays, including research on hormesis, isopathy and similitude.

Thus, maybe, a good way to validate a physical-chemical model or theoretical proposition is correlating it to biological experiments. If one observes changes on a physical property due to potentization, the only way to associate these observations with the homeopathic phenomenon is correlating them with changes on biological responses. Otherwise, one will always be in doubt about what one really is observing: experimental artifacts, unexpected properties or the natural aqueous solution dynamics.

Conclusion

Physical-chemical experiments performed without biological correspondence cannot contribute significantly to the understanding of homeopathic phenomena itself, since several experimental evidences point toward the necessity of the presence of a living system. Thus, one can not neglect this when proposing an experimental methodology. Multidisciplinary studies should be a common practice in the scientific homeopathic community.

Also, theoretical models need to be built keeping in mind the biological interaction. Entangled states or non locality can be useful concepts to open the mind; evidences about the quantum nature of the homeopathic phenomenon are still lacking.

Molecular structural theoretical models isolated from biological evidences are inadequate to explain the phenomenon. Phenomenological and informational models have, in principle, the advantage to be independent of molecular structure. However, these can be considered only interpretative, describing only some experimental features, but never giving them any explanation. Also, the nature of the so called information and their principles of action must be established.

In conclusion, experimental or theoretical models must incorporate the fundamental characteristic of the homeopathic phenomenon: it happens only in the presence of a life system. To ignore such characteristics is a risk ignoring its validity limits.

References

Aabel S, Fossheim S, Rise F. Nuclear magnetic resonance (NMR) studies of homeopathic solutions. Br. Hom. J.; 2001, 90: 14–20.

Anagnostatos GS. Small water clusters (clathates) in the preparation process of Homeopathy. In: Endler PC, Schulte J (Eds). Ultra High Dilution: Physiology and Physics. Dordrecht: Kluwer ; 1994.

Anick DJ, Ives JA. The silica hypothesis for homeopathy: Physical chemistry. Homeopathy; 2007, 96: 189–195.
Atmanspacher H, Römer H, Wallach H. Weak quantum theory: Complementary and entanglement in physics and beyond. Found Physics; 2002, 32: 379–406.
Bastide M, Doucet-Jaboeuf M, Daurat V. Activity and chronopharmacology of very low doses of physiological immune inducers. Immunol. Today; 1985, 6: 234–235.
Bastide M, Daurat V, Doucet-Jaboeuf M, Pelegrin A, Dorfman P. Immunomodulatory activity of very low doses of thymulin in mice. Int. J. Immunother.; 1987, 3: 191–200.
Bastide M, Lagache A, Lemaire-Misonne C. Le paradigme des signifiants: Schème d'information applicable à l'Immunologie et à l'Homéopathie. Revue Intern. Systémique; 1995, 9: 237–249.
Betti L, Brizzi M, Nani D. Effect of high dilutions of arsenicum album on wheat seedlings from seed poisoned with the same substance. Br. Hom. J.; 1997, 86: 86–89.
Bonamin LV, Nina AL, Caviglia F, Martinho K. Very low dilutions of dexamethasone inhibit their own pharmacological effect *in vivo*. Br. Hom. J.; 2001, 90: 198–203.
Brizzi M, Nani D, Peruzzi M, Betti L. Statistical analysis of the effect of high dilutions of arsenic in a large dataset from a wheat germination model. Br. Hom. J.; 2000, 89: 63–67.
Chaplin MF. The memory of water: An overview. Homeopathy; 2007, 96: 143–150.
Dantas F, Rampes H. Do homeopathic medicines provoke adverse effects? A systematic review. Br. Hom. J.; 2000, 89(Suppl. 1): S35–338.
Dantas F, Fisher P, Walach H, Wieland F, Rastogi DP, Teixeira H, Koster D, Jansen JP, Eizayaga J, Alvarez MEP, Marim M, Belon P, Weckx LLM. A systematic review of the quality of homeopathic pathogenetic trials published from 1945 to 1995. Homeopathy; 2007, 96: 4–16.
Demangeat JL, Demangeat C, Gries P, Poitevin B, Constantinesco A. Modifications des remps de relaxation RMN à 4 MHz des protons du solvant dans les très hautes dilutions salines de silice/lactose. J. Med. Nucl. Biophys.; 1992, 16: 135–145.
Elia V, Marchese M, Montanino M, Napoli E, Niccoli M, Nonatelli L, Ramaglia A. Hydrohysteretic phenomena of extremely diluted solutions induced by mechanical treatments: A calorimetric and conductometric study at 25°C. J. Solution Chem.; 2005, 34: 947–960.
Elia V, Elia L, Cacace P, Napoli E, Noccoli M, Savarese F. The extremely diluted solutions as multi-variable systems: A study of calorimetric and conductometric behaviour as a function of the parameter time. J. Therm. Anal. Calorimet. 2006, 84: 317–323.
Elia V, Napoli E, Germano R. The memory of water: An almost deciphered enigma. Dissipative structures in extremely dilute aqueous solutions. Homeopathy; 2007, 96: 163–169.
Endler PC, Pongratz W, Kastberger G, Wiegant FAC, Schulte J. The effect of highly diluted agitated thyroxine on the climbing activity of frogs. J. Vet. Hum. Tox.; 1994, 36: 56–59.
Endler PC, Pongratz W, Smith CW, Schulte J. Non-molecular information transfer from thyroxine to frogs with regard to homeopathic toxicology. Vet. Hum. Tox.; 1995, 37: 259–260.
Endler PC, Pongratz W, Smith CW, Schulte J, Senekowitsch F, Citro M. Non molecular information transfer from thyroxine to frogs. In: Bastide M. (Ed). Signal and Images. Dordrecht: Kluwer; 1997.
Jurj G. Signs in the medical context and the semiotic codes: An introduction to medical semiotics. Cultura Homeopática; 2007, 20: 32–36.
Klimek W, Yan Y, Tschulakow AV. A new approach to the memory of water. Homeopathy; 2005, 94: 241–247.
Lagache A. What is Information ? In: Bastide M (Ed). Signal and Images. Dordrecht: Kluwer; 1997a.
Lagache A. Notes on the conceptual basis of Science. In: Bastide M (Ed). Signal and Images. Dordrecht: Kluwer; 1997b.
Lobyshev VI, Shikhlinskaya RE, Ryzhikov BD. Experimental evidence for intrinsic luminescence of water. J. Mol. Liq.; 1999, 82: 73–81.
Milgrom LR. Conspicuous by its absence: The memoty of water, macro-entanglement and the possibility of homeopathy. Homeopathy; 2007, 96: 209–219.

Oberbaum M, Vithoulkas G, Hasalen RV. Clinical Trials of Classical Homeopathy: Reflections on Appropriate Research Designs. J. Alter. Complem. Med.; 2003, 9(1): 105–111.
Prigogine I. Time, Structure and Fluctuations. Science; 1978, 201: 777–785.
Rey L. Thermoluminescence of ultra high dilutions of lithium chloride and sodium chloride. Physica A; 2003, 323: 67–74.
Rey L. Can low-temperature thermoluminescence cast light on the nature of ultra-high dilutions? Homeopathy; 2007, 96: 170–174.
Samal S, Geckeler KE. Unexpected solute aggregation in water on dilution. Chem. Commun.; 2001, 21: 2224–2225.
Singer S, Amit-Kohn M, Weiss S, Rosenblum J, Lukasiewicz E, Itzchaki M, Oberbaum M. Efficacy of a homeopathic preparation in control of pos-operativa pain – A pilot clinical trial. Acute Pain; 2007, 9: 7–12.
Teixeira J. Can water possibly have a memory? A skeptical view. Homeopathy; 2007, 96: 158–162.
Van Wijk R, Bosman S, van Wijk EP. Thermoluminescence in ultra-high dilution research. J. Alter. Complem. Med.; 2006, 12(5): 437–443.
Weingärtner O. The nature of the active ingredient in ultramolecular dilutions. Homeopathy; 2007, 96: 220–226.
Youbicier-Simo BJ, Boudard F, Mekaouche M, Bastide M, Baylé JD. Effects of embryonic bursectomy and *in ovo* administration of highly diluted bursin on adrenocorticotropic and immune response of chickens. Int. J. Immunotherap.; 1993, 9: 169–180.
Youbicier-Simo BJ, Boudard F, Mekaouche M, Baylé JD, Bastide M. A role for Bursa Fabricii and bursin in the ontogeny of the pineal biosynthetic activity in the chicken. J. Pineal Res.; 1996a, 21: 35–43.
Youbicier-Simo BJ, Boudard F, Mekaouche M, Baylé JD, Bastide M. Specific abolition reversal of pituitary-adrenal activity and control of the humoral immunity in bursectomized chickens through highly tern. J. Immunopathol. Pharmacol.; 1996b, 9: 43–51.

Chapter 3
Mechanical Versus Handmade Succussions: A Physical Chemistry Comparison

Carla Holandino[1], Felipe Dias Leal[1], Bianca de Oliveira Barcellos[1], Maria Augusta Campos[1], Raíza Oliveira[1], Venício Feo da Veiga[1], Sheila Garcia[1], and Carlos Renato Zacharias[2]

Introduction

Homeopathic medicines are prepared all over the world by a serial process of dilution and shaking (*succussion*), named dynamization or potentization. The dilution is generally performed as decimal or centesimal (in volume) while *succussion* can be made by mechanical, handmade, vortex, among other procedures (Fontes, 2005; Martinez, 1990). The proposition of each one is based on cultural, technological, commercial or philosophical reasoning. Experimental results have demonstrated the efficacy of all these procedures, validating the different *succussion* techniques (Belon et al., 2004; Davenas et al., 1988; Bonamin et al., 2001; Varricchio et al., 2006).

The most common way to perform *succussions* is by keeping a liquid preparation inside a glass vessel and shaking it violently using a mechanical apparatus or beating it against a hard, but elastic surface in handmade work. Handmade *succussions* were proposed by Hahnemann (2003), probably inspired on alchemist techniques (Ruiz, 2002). Mechanical *succussion* had its origin probably on Industrial Revolution, with the desire to produce higher potencies faster, cheaper and in a standardized way (Wiston, 1989), and also because of the belief that the emotional state of the manipulator could affect the medicines. An important feature of mechanical *succussion* is the intense production of bubbles on the liquid phase. Physical chemistry reasoning leads us to conclude that bubbles increase the superficial contact area and gasification, thus, some chemical degradation process occurs. However, no one knows if and how this compounding procedure affects the efficacy of homeopathic medicines. Also, *succussion* can be interpreted as a mechanism that allows transfer of the mechanical energy input all the way down to the molecular level, where it becomes available to perform chemical work (Torres, 2002; Anagnostatos, 1994; Auerbach, 1994).

[1]Universidade Federal do Rio de Janeiro, Rio de Janeiro, Brazil. cholandino@yahoo.com

[2]Universidade Estadual Paulista, Guaratinguetá, Brazil
Presented at the XXth Symposium of GIRI, São Paulo, 2006

Which procedure is more efficient? This point still remains open and the homeopathic pharmacopoeia just regulates the procedure of both *succussion* techniques, without compare them (*Farmacopéa Homeopática Brasileira,* 1997).

This work was proposed with the aim of shedding some light on this point, using scientific experimental methods. To address this issue, one first has to think about what "efficient" means. The best way to define it is through clinical results or proving trials. However, this approach is not within the scope of this chapter. So, we looked for a parameter able to reflect some physical chemistry medicine properties, which could be used as a marker.

Recently, some papers were published describing the physical chemistry properties of extremely diluted aqueous solutions (Elia et al., 2004, 2005, 2006, 2007; Rey, 2003, 2007; Van Wijk et al., 2006; Lobyshev et al., 1999; Samal and Geckeler, 2001; Walach et al., 1988; Klimek et al., 2005; Rao et al., 2007). From these, the electrical conductivity of solutions seemed to be an adequate indicator of structural changes induced by potentization processes, and thus, might furnish some relevant information for the comparison of *succussion* techniques (Elia et al., 2005, 2006, 2007).

The aim of this paper is to compare mechanical and handmade preparations based on electrical conductivity measurements, using the glasses and the procedures commonly used by pharmacies to prepare homeopathic remedies. For that purpose, glasses type III were used, a mechanical machine (AUTIC 10–50) and a conductivity apparatus (Metler-Toledo MPC 227). The main hypothesis is that one can find a mathematical correlation between mechanical and handmade preparations, establishing a comparative criterion based on the electrical conductivity measurements.

Material and Methods

The solutions were prepared with Vincristine sulfate – VCR – (Zodiac®) on purified water (Millipore®) using amber glasses (USP type III soda-lime glass – DIN 168) with a plastic cover used under the caps to seal the flasks. This classification was determined by USP criteria (United States Pharmacopoeia, 2007) based on the following characteristics: Type III glass is suitable for packing liquid formulations that are insensitive to alkali and it also has the advantage that it could be used in dry heat sterilization. Such type of glass is the most common one, used by pharmacies, to prepare homeopathic medicines.

The starting solution consisted of 1.0 mg/ml VCR diluted in 10 ml of purified water, in a 15 ml glass. Centesimal hahnemannian potencies were produced until 15 CH (potentized VCR). Mechanical *succussions* (Denise 10–50, Autic®) were performed with 100 *succussions* for 33 seconds (approximately 3 Hz). In order to register some eventual chemical effect of VCR or systematic experimental errors on potencies, an equivalent set of control samples (1–15 CH) was prepared with purified water only (potentized water).

Electrical conductivity measurements were performed on all samples at 25°C with systematic calibration and temperature compensation using a Mettler-Toledo MPC 227 apparatus. Intrinsic experimental errors were 0.5%. The data from the samples were collected immediately after sample preparation (time zero) and after 7, 14, 21 and 35 days. The glasses were kept on rest in the intervals. Each sample was measured 4 (four) times and the experiment was repeated twice independently to evaluate the average value and standard deviation.

Results

Tables 3.1–3.4 report the experimental data (electrical conductivity) for VCR and water potencies, prepared mechanically (X_m) and in a handmade manner (X_h). All electrical conductivity values are on μS/cm and time in days. The data are presented as averaged value with standard deviation.

Tables 3.1 and 3.2 refer to mechanically *succussion*ed samples, while 3 and 4, to handmade ones. As our work hypothesis is the existence of a mathematical correlation between both techniques, one must compare Tables 3.1 and 3.3 for VCR, as well Tables 3.2 and 3.4 for water potencies. It can be done graphically rearranging the data, as showed on Table 3.5.

Rearranging the data as on Table 3.5, any temporal effect was eliminated over the samples, due to water chemical dynamic effects. Also, assuming that these temporal effects act similarly on both sets, arranging all pairs of data in the same

Table 3.1 Electrical conductivity values for VCR – mechanical succusions (X_m)

t	1CH	4CH	7CH	9CH	12CH	13CH	14CH	15CH
0	18.67 ± 0.98	6.9 ± 2.9	4.9 ± 1.3	6.7 ± 2.9	4.9 ± 1.4	6.1 ± 1.5	5.95 ± 0.87	4.71 ± 0.48
7	18.6 ± 1.5	9.0 ± 3.2	5.89 ± 0.77	9.3 ± 2.2	6.7 ± 1.0	10.02 ± 0.83	9.3 ± 1.6	8.2 ± 1.3
14	20.04 ± 0.97	10.7 ± 3.4	7.4 ± 1.4	10.2 ± 3.6	8.4 ± 1.4	9.84 ± 0.71	9.5 ± 1.4	9.0 ± 1.2
21	21.08 ± 0.71	12.1 ± 3.2	9.4 ± 1.7	11.3 ± 3.8	9.7 ± 1.4	10.8 ± 1.1	10.3 ± 1.3	11.00 ± 0.79
35	22.5 ± 1.8	15.9 ± 2.6	14.2 ± 5.1	13.8 ± 4.4	11.6 ± 1.5	13.5 ± 2.4	15.0 ± 3.5	12.71 ± 0.33

Table 3.2 Electrical conductivity values for water – mechanical succusions (X_m)

t	1CH	4CH	7CH	9CH	12CH	13CH	14CH	15CH
0	4.4 ± 1.4	5.1 ± 2.1	5.4 ± 1.9	7.1 ± 3.2	5.2 ± 1.4	8.8 ± 2.8	6.7 ± 1.5	4.73 ± 0.59
7	6.1 ± 1.4	7.8 ± 2.5	7.4 ± 3.8	8.4 ± 2.8	8.1 ± 1.6	10.1 ± 2.0	9.6 ± 1.6	7.98 ± 0.88
14	8.5 ± 3.4	9.2 ± 2.0	8.0 ± 3.6	9.8 ± 2.6	8.49 ± 0.71	12.3 ± 2.6	10.6 ± 1.2	8.22 ± 0.14
21	10.7 ± 3.3	10.7 ± 1.9	10.0 ± 4.4	11.9 ± 2.8	10.5 ± 1.6	13.7 ± 3.0	13.9 ± 1.4	9.71 ± 0.08
35	13.6 ± 3.7	13.2 ± 1.3	12.2 ± 4.5	14.0 ± 3.0	12.8 ± 2.0	14.8 ± 3.2	17.0 ± 1.7	11.9 ± 1.2

Table 3.3 Electrical conductivity values for VCR – handmade succusions (X_h)

t	1CH	4CH	7CH	9CH	12CH	13CH	14CH	15CH
0	18.0 ± 3.6	3.91 ± 0.03	4.14 ± 0.91	8.6 ± 3.6	5.55 ± 0.86	5.49 ± 0.54	5.40 ± 0.16	3.80 ± 0.11
7	18.6 ± 4.3	6.16 ± 0.23	5.97 ± 0.97	8.6 ± 1.7	6.58 ± 0.76	7.08 ± 0.21	6.14 ± 0.05	5.50 ± 0.08
14	21.4 ± 7.8	10.1 ± 2.1	10.8 ± 5.6	12.6 ± 5.9	8.03 ± 0.82	9.60 ± 0.52	7.46 ± 0.15	7.60 ± 0.40
21	26 ± 11	11.8 ± 1.7	13.0 ± 6.7	14.1 ± 5.4	10.73 ± 0.83	11.41 ± 0.06	8.49 ± 0.30	9.60 ± 0.59
35	29 ± 14	13.5 ± 1.0	16.9 ± 7.4	16.5 ± 5.2	12.73 ± 0.25	13.27 ± 0.20	10.93 ± 0.37	11.5 ± 1.1

Table 3.4 Electrical conductivity values for water – handmade succusions (X_h)

t	1CH	4CH	7CH	9CH	12CH	13CH	14CH	15CH
0	18.0 ± 3.6	3.91 ± 0.03	4.14 ± 0.91	8.6 ± 3.6	5.55 ± 0.86	5.49 ± 0.54	5.40 ± 0.16	3.80 ± 0.11
0	4.6 ± 1.3	3.97 ± 0.03	4.09 ± 0.11	6.0 ± 1.3	4.6 ± 1.6	5.4 ± 1.7	5.03 ± 0.71	5.0 ± 1.6
7	5.5 ± 1.3	5.32 ± 0.57	5.89 ± 0.16	8.3 ± 1.9	6.9 ± 1.1	7.2 ± 2.1	6.71 ± 0.48	6.14 ± 0.88
14	7.5 ± 3.1	7.8 ± 2.6	7.30 ± 0.44	10.0 ± 2.6	8.6 ± 1.0	8.7 ± 2.1	8.48 ± 0.41	7.73 ± 0.72
21	9.4 ± 1.3	9.2 ± 2.8	8.78 ± 0.73	12.0 ± 3.9	9.75 ± 0.63	10.8 ± 2.0	10.03 ± 0.20	9.25 ± 0.86
35	11.22 ± 0.56	11.4 ± 3.5	11.3 ± 1.4	13.7 ± 3.8	11.61 ± 0.66	13.1 ± 2.6	12.53 ± 0.76	11.4 ± 1.0

Table 3.5 Rearranged data codes. A table element is described as nCH – t_m. It represents the electrical conductivity of the potency *n* CH, measured at day *m*. Column at left represents the handmade samples, while at right, the mechanical ones. Using this codification, one can form two new tables, corresponding to VCR and water data. The complete Table 3.5 for VCR (Table 3.7) and water potencies (Table 3.8) can be found in appendix A

Sample	
Handmade (X_h) (Tables 3.3 or 3.4)	Mechanical (X_m) (Tables 3.1 or 3.2)
Data: n CH – t_m (µS/cm – day)	
1CH – t_0	1CH – t_0
…	…
15 CH – t_0	15 CH – t_0
1CH – t_7	1CH – t_7
…	…
15 CH – t_7	15 CH – t_7
1CH – t_{14}	1CH – t_{14}
…	…
…	…
15 CH – t_{35}	15 CH – t_{35}

table must generate a smoothed curve, with many experimental points able to perform better numerical analyses.

In this way, two new tables were obtained, whose graphical representations are shown in Figs. 3.1 and 3.2. The axes contain the electrical conductivity value for the samples, with handmade ones (X_h) on the horizontal axis and mechanically (X_m), on the vertical axes. The propagated standard deviation is represented as bar deviations. The linear fitting is also included but the coefficients are discussed in the next section.

Both figures show a linear correlation between mechanical and handmade preparations. Thus, linear regression analyses were performed to calculate the angular (A), linear (B) and correlation (r^2) coefficients, reported on Table 3.6.

Discussion

In this paper, it was proposed to compare mechanical and handmade preparations based on electrical conductivity measurements, using glasses (Type III - DIN 168) and procedures commonly used by pharmacies to prepare homeopathic remedies,

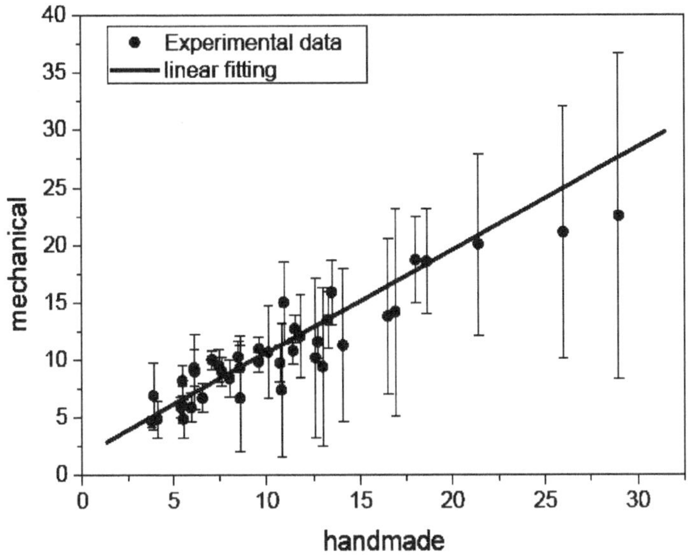

Fig. 3.1 Experimental data and linear fitting for VCR: $X_m = (0{,}894 \pm 0.077) * X_h + (1.70 \pm 0.61)$; $R^2 = 0.93$; $p < 0.0001$

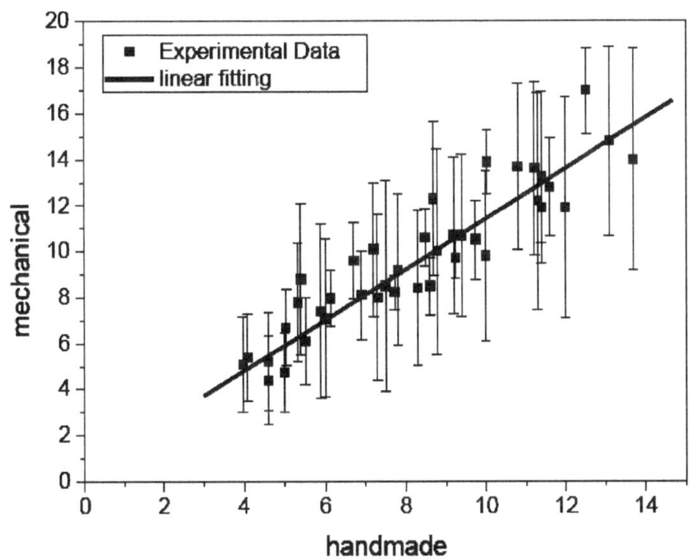

Fig. 3.2 Experimental data and linear fitting for water: $X_m = (1.10 \pm 0.14) * X_h + (0.4 \pm 1.1)$; $R^2 = 0.90$; $p < 0.0001$

in order to study a real situation. Glass types can be described accordingly to their characteristics, as follows: (1) Type I glass designed for use in all products requiring very high resistance to strong acids or alkalis and in products intended for use in heat applications such as autoclaves. Its chemical composition is a borosilicate (80.6%) with traces of oxides (0.1% of K_2O). (2) Type II glass is formed as a borosilicate (96.4%), but it is subjected to a chemical treatment that removes most of the elements in the glass, except silica (Si_2O). It is designed for use in all products that must withstand very high temperatures or thermal shock. (3) Type III glass is a brand borosilicate glass (72.3%) plus alkaline oxides, such as, Na_2O (8.6%), Al_2O_3 (5.9%), CaO (0.8%), K_2O (1.2%), BaO (2.5%) and others. Most homeopathic pharmacies have been choosing Type III glass for the compounding of homeopathy remedies because this kind of glass presents a higher mechanical resistance (the glasses are submitted to repeated shock in the dynamization procedure) and a moderate hydrolytic resistance (medicines are essentially water and ethanol) [USP 28]. Also, one may consider the economical point, which is quite important for commercial pharmacies, since this kind of glass meets the quality parameters established by the homeopathic pharmacopoeia and are reasonably priced.

Although Type III glass releases some contaminants in the liquid phase, clinical effects are observed, which means that clinical effects are not invalidated by such contaminants. Herein the use of such kind of glasses was correct and desirable, because they are present on the daily clinical practice. Anyway, there was always a parallel control to compare results. Further, Elia et al. recently showed that the presence of impurities released by the glassware contributes to the physicochemical state of the dilutions, but it is not relevant in comparison with the auto-organization process of the water molecules (Elia et al., 2007).

The analysis of Table 3.6 can start with the correlation coefficient (r^2). It indicates the fitting quality of the experimental data by a linear function. The obtained values (0.93 and 0.90) mean that linear correlation is a good model, reinforcing the composition of Table 3.5. Thus, one can establish that:

$$X_m = A^*Xh + B$$

Since the linear correlation is validated, one must interpret the linear coefficient (B). The calculated value for water potencies (Fig. 3.2) is close to zero, indicating that both *succussion* procedures have similar effects on the water potencies. However, for VCR-data (Fig. 3.1), such coefficient is close to 1.70 μS/cm. This value may be caused by 2 factors: (i) the propagated experimental uncertainties; and (ii) the presence of VCR's molecules on 1–4 cH samples. The first factor would also affect the water potencies and can be estimated as about 0.5 (μS/cm), so not enough to explain alone the value of the linear coefficient.

Table 3.6 Mathcmatical linear model: $X_m = A * X_h + B$; R2 is the correlation coefficient and p-value is the t-test significance level

	VCR	Water
A	0.894 ± 0.077	1.10 ± 0.14
B	1.70 ± 0.61	0.4 ± 1.1
R^2	0.93	0.90
p-value	<0.0001	<0.0001

The second factor seems to be more significant in our experiments. Vincristine is a cytostatic substance of the Vinca-alcaloides family which is often used in chemotherapy (Sedlecki et al., 1989). In this study, any chemical degradation induced by the *succussion* or temperature is discarded, since this drug has a good stability in our experimental conditions. However, the presence of molecules of VCR in the 1 CH potency can affect the linear fitting making it tend, to a non-zero linear coefficient (see higher electric conductivity points in Fig. 3.1). Such perturbation is not observed in water samples.

Also, the quantity of bubbles formed on mechanical techniques exceeds significantly those from handmade one. As the bubbles increase the superficial contact area and gasification, affecting strongly the liquid dynamics, a more intense chemical degradation might be supposed, resulting on ion rearrangement, changing the electrical conductivity. It might happen only for the earlier potencies (<4 CH), while the active substance (VCR) is present. Rey (2007) recently showed that mechanical *succussion* can change the thermoluminescence glow of ultra-high dilutions (UHD) of LiCl, NaCl and D_2O (15 CH), associating such effects to the formation of nanobubles (Rey, 2007). The intense presence of bubbles increases the gas-water interface affecting the specific distributions of ions in water at the boundary of these bubbles (Vallée et al., 2005). However, if this affects the efficacy of homeopathic medicine, it remains an open question. In fact, the number of bubbles produced by mechanical *succussion* machines is higher than those from manual *succussions*, which might affect the chemical stability and dynamics of UHD especially for low potencies (1–4 CH). Such hypothesis must be confirmed because its consequences are very important as they affect the chemical integrity of the active substance and, probably, its clinical effects.

The interpretation of the angular coefficient (A) is the most interesting, and in the present work, the most important. Both data sets (VCR and water) present A-value close to 1.0. It means that, within the experimental uncertainties and limitations, mechanical and handmade *succussions* are equivalent, or it is not possible to distinguish, in terms of electrical conductivity, the medicines prepared by one or the other procedure.

Such result leads us to conclude that bubble formation has no relation electrical conductivity except, probably, for the very low dilutions (<6 CH). Thus, with the increase of the reactive surface, turbulence, cavitations, gasification and any other bubble-related phenomena seem to have no relevance in the potentization process. Also, it might reinforce the use of vortex procedures, but this discussion is not the purpose of this work.

However, another interpretation for the unitary angular coefficient can be done, inquiring about the electric conductivity dependence on the potentization process and its classification as a good indicator parameter for the comparisons.

Recently, some papers were published describing physical chemistry properties of extremely diluted aqueous solutions (Elia et al., 2004, 2005, 2006, 2007; Rey, 2003, 2007; Van Wijk et al., 2006; Lobyshev et al., 1999; Samal and Geckeler, 2001; Walach et al., 1988; Klimek et al., 2005; Rao et al., 2007). Some of them (Elia et al., 2004, 2005, 2006, 2007) showed that the electrical conductivity of extremely diluted solutions seemed to be an adequate indicator of structural changes induced by potentization process, and thus, might present relevant information for *succussion* techniques comparison. Besides electrical conductivity, Elia et al. (2007) showed, using other physicochemical methodologies, such as pHmetry, flux calorimetry and galvanic cell electrode potential, that the ageing of the homeopathic solutions modifies the physicochemical nature of these solutions.

From the present data, it is not possible to draw any conclusions about clinical results or action mechanisms. But, supposing that electrical conductivity has no relation with the potency, the obtained data would be inadequate to distinguish among *succussion* procedures. Moreover, follows reasoning, the idea of using physical-chemical parameters to investigate the homeopathic phenomenon might be inadequate to understand such phenomenon.

This last conclusion is in complete disagreement with our initial argumentation of using electrical conductivity for compare *succussions* procedures, and also with some recent publications. However, it can not be discharged without further analyses.

Conclusion

This study produced two uncomfortable conclusions: mechanical and handmade preparations are equivalent or, electric conductivity is not a good parameter to distinguish them.

It is important to go deeper inside them to realize theirs consequences. If the first conclusion is valid, the relevance of bubbles and other hydrodynamics effects on potentization procedure must be questioned. Bubbles, cavitations and turbulence effects are involved on energy transfer, water clusters formation and others chemical and dynamical effects. For the low dilutions, in which molecules from the starting sample (mother tincture) can be found, such effects could induce chemical reactions and degradation of some molecules, probably reflecting on clinical results. The present experimental data did not show any significant difference between mechanical and handmade preparations in terms of electrical conductivity to the tested substance (VCR), except for lower dilutions (<4CH). If the liquid dynamics does not affect the medicine preparation, the use of vortex, continuous flux or any other method to perform *succussions* of homeopathic remedies could be considered as efficient methods. Further, one could suppose that the solvent dynamics does not exert any rule on homeopathic phenomenon. Of course, this first conclusion must be formerly checked by submission to others experimental techniques.

The second conclusion put in check the validity of electrical conductivity as a good parameter to study homeopathic potencies. If this conclusion is valid, any correlation between *succussion* procedures could not be established, except the trivial linear correlation with unitary angular coefficient. The linear coefficient would just reflect experimental uncertainties or some VCR residual influence. Such conclusion leads to the question that electrical conductivity could not change with potencies, in disagreement with other authors and even with the present experimental data. Also, if the physical-chemical changes are associated with water structural rearrangement, as pointed out in the literature, no structural changes related to potentization could happen. Our graphical analyses (Figs. 3.1 and 3.2) consisted of distributing the data from each *succussion* procedure on separated axis and all common factors or errors affecting equally both procedures were eliminated. But, if these factors really exist and act equally, they are not involved in the homeopathic phenomenon or are not relevant for potencies comparisons. This second conclusion is uncomfortable, because it works against structural models for *potentized* (or dynamized) systems, requiring new reasoning or models to explain the phenomenon. Thus, this second conclusion must be also revisited.

Both conclusions must be put in check with other controlled experimental techniques, because they point against solvent structural models or the existence of unexpected water properties. Also, opens the mind to propose different techniques for *succussion*, while inquiring about chemical changes on active substances. Finally, put some doubt over the reliability of physical chemistry experiments, performed directly on medicines, without a biological sensor involved.

Acknowledgements We thank Dr Mauro Sola-Penna for critical reading of the manuscript. This work was supported by Conselho Nacional de Desenvolvimento Científico e Tecnológico (CNPq), Fundação de Amparo a Pesquisa no Estado do Rio de Janeiro (FAPERJ), Fundação José Bonifácio (FUJB), Fundação Ary Franzino/Fundação Educacional Charles Darwin (FAF/FECD/ONCO) and Projetos de Pesquisa para o Sistema Único de Saúde (PPSUS/FAPERJ).

References

Anagnostatos GS. Small water clusters (clathates) in the preparation process of Homeopathy. In: Endler PC, Schulte J (Eds). Ultra High Dilution: Physiology and Physics. Dordrecht: Kluwer; 1994.

Auerbach D. Mass, fluid and wave theory motion during the preparation of the ultra high dilutions. In: Endler PC, Schulte J (Eds). Ultra High Dilution: Physiology and Physics. Dordrecht: Kluwer; 1994.

Belon P, Cumps J, Ennis M, Mannaioni PF, Roberfroid M, Sainte-Laudy J, Wiegant FAC. Histamine dilutions modulate basophil activation. Inflamm. Res.; 2004, 53: 181–188.

Bonamin LV, Martinho KS, Nina AL, Caviglia F, Do Rio RGW. Very high dilutions of Dexamethasone inhibit their pharmacological effect *in vivo*. Brit. Hom. J.; 2001, 90: 198–203.

Davenas E, Beauvais F, Amara J, Oberbaum M, Robinzon B, Miadonna A, Tedeschi A, Pomeranz B, Fortner P, Belon P, Sainte-Laudy J, Poitevin B, Benveniste J. Human basophil degranulation triggered by very dilute antiserum against IgE. Nature; 1988, 333: 816–818.

Elia V, Baiano S, Duro I, Napoli E, Niccoli M, Nonatelli L. Permanent physical chemistry properties of extremely diluted aqueous solutions of homeopathic medicines. Homeopathy; 2004, 93: 144–150.

Elia V, Marchese M, Montanino M, Napoli E, Niccoli M, Nonatelli L, Ramaglia A. Hydrohysteretic phenomena of extremely diluted solutions induced by mechanical treatments: A calorimetric and conductometric study at 25°C. J. Sol. Chem.; 2005, 34: 947–960.

Elia V, Elia L, Cacace P, Napoli E, Noccoli M, Savarese F. The extremely diluted solutions as multi-variable systems: A study of calorimetric and conductometric behaviour as a function of the parameter time. J. Therm. Anal. Calorimet.; 2006, 84: 317–323.

Elia V, Napoli E, Germano R. The "Memory of Water": An almost deciphered enigma. Dissipative structures in extremely dilute aqueous solutions. Homeopathy; 2007, 96: 163–169.

Farmacopéia Homeopática Brasileira, 2ª Ed. São Paulo: Atheneu; 1997.
Fontes OL. Farmácia Homeopática. 2ª Ed. São Paulo: Manole; 2005.
Hahnemann S. Organon of Medicine, 6th Ed. London: Orion Books; 2003.
Klimek W, Yan Y, Tschulakow AV. A new approach to the memory of water. Homeopathy; 2005, 94: 241–247.
Lobyshev VI, Shikhlinskaya RE, Ryzhikov BD. Experimental evidence for intrinsic luminescence of water. J. Mol. Liq.; 1999, 82: 73–81.
Martinez J. Dinamizacion. In: A. Farmacia Homeopática Ed. Buenos Aires: Albators; 1990.
Rao ML, Roy R, Bell IR, Hoover R. The defining role of structure (including eptiaxy) in the plausibility of homeopathy. Homeopathy; 2007, 96: 175–182.
Rey L. Thermoluminescence of ultra high dilutions of lithium chloride and sodium chloride. Physica A; 2003, 323: 67–74.
Rey L. Can low-temperature thermoluminescence cast light on the nature of ultra-high dilutions? Homeopathy; 2007, 96: 170–174.
Ruiz R. Da Alquimia à Homeopatia. Bauru: Edusc; 2002.
Samal S, Geckeler KE. Unexpected solute aggregation in water on dilution. Chem. Commun.; 2001, 21: 2224–2225.
Sedlecki S, Jakovljevic A, Pap K. Effect of vincristine on molar development and jaw growth in rats. Stomatol Glas Srb.; 1989, 36(4): 289–297.
Torres JL. Physical basis of *succussion*. Homeopathy; 2002, 91: 221–224.
United States Pharmacopoeia. 30Ed. Rockville: The United States Convention; 2007.
Vallée P, Lafait J, Legrand L, Mentré P, Monod MO, Thomas Y. Effects of pulsed low-frequency electromagnetic fields on water characterized by light scattering techniques: Role of bubbles. Langmuir; 2005, 21: 2293–2299.
Van Wijk R, Bosman S, van Wijk EP. Thermoluminescence in ultra-high dilution research. J. Altern. Complement. Med.; 2006, 12(5): 437–443.
Varricchio MCBN, Aquino CL, Moreno GB, Pereira FG, Veiga VF, Zancan P, Sola-Penna M, Kuster RM, Holandino C. Cytotoxicity and antitumoral activity of extremely diluted solutions of *Euphorbia tirucalli Lineu*: An *in vitro* study. Improving the Success of Homeopathy: A global perspective (5). 2006; London.
Walach H, Assendonk T, Bourkas P, Delininick A, Ives G, Karraginnopoulos C, Lüdtke R, Wassenhoven M, Witt C. Electric measurement of ultra-high dilutions – a blinded controlled experiment. Brit. Hom. J.; 1988, 87: 3–12.
Wiston J. A brief history of potentizing machines. Brit. Hom. J.; 1989, 78: 59–68.

Appendix

Table 3.7 Reorganized data for VCR samples. Columns sequence: handmade (X_h), mechanical (X_m), propagated experimental deviation

X_h	X_m	Exp. Deviation
18	18.67	3.73
3.91	6.9	2.9
4.14	4.9	1.59
8.6	6.7	4.62
5.55	4.9	1.64
5.49	6.1	1.59
5.4	5.95	0.88
3.8	4.71	0.49
18.6	18.56	4.55
6.16	9	3.21
5.97	5.89	1.24
8.6	9.3	2.78
6.58	6.7	1.26
7.08	10.02	0.86
6.14	9.3	1.6
5.5	8.2	1.3
21.4	20.04	7.86
10.1	10.7	4
10.8	7.4	5.77
12.6	10.2	6.91
8.03	8.4	1.62
9.6	9.84	0.88
7.46	9.5	1.41
7.6	9	1.26
26	21.08	11.02
11.8	12.1	3.62
13	9.4	6.91
14.1	11.3	6.6
10.73	9.7	1.63
11.41	10.8	1.1
8.49	10.3	1.33
9.6	11	0.99
29	22.5	14.12
13.5	15.9	2.79
16.9	14.2	8.99
16.5	13.8	6.81
12.73	11.6	1.52
13.27	13.5	2.41
10.93	15	3.52
11.5	12.71	1.15

Table 3.8 Reorganized data for water samples. Columns sequence: handmade (X_h), mechanical (X_m), propagated experimental deviation

X_h	X_m	Exp. deviation
4.6	4.4	1.91
3.97	5.1	2.1
4.09	5.4	1.9
6	7.1	3.45
4.6	5.2	2.13
5.4	8.8	3.28
5.03	6.7	1.66
5	4.73	1.71
5.5	6.1	1.91
5.32	7.8	2.56
5.89	7.4	3.8
8.3	8.4	3.38
6.9	8.1	1.94
7.2	10.1	2.9
6.71	9.6	1.67
6.14	7.98	1.24
7.5	8.5	4.6
7.8	9.2	3.28
7.3	8	3.63
10	9.8	3.68
8.6	8.49	1.23
8.7	12.3	3.34
8.48	10.6	1.27
7.73	8.22	0.73
9.4	10.7	3.55
9.2	10.7	3.38
8.78	10	4.46
12	11.9	4.8
9.75	10.5	1.72
10.8	13.7	3.61
10.03	13.9	1.41
9.25	9.71	0.86
11.22	13.6	3.74
11.4	13.2	3.73
11.3	12.2	4.71
13.7	14	4.84
11.61	12.8	2.11
13.1	14.8	4.12
12.53	17	1.86
11.4	11.9	1.56

Chapter 4
Water and High Dilutions Phenomenology: Physical Characterization

Adriana Ramos de Miranda

Introduction

For several decades the study of water behavior has motivated the work of a great number of researchers from different areas of knowledge, such as chemistry, biology, medicine, in addition to the specific areas concerned with human behavior and social development. Physics has contributed for a long time to the comprehension of phenomena that seem to be irremediably dependent of water.

In the last decades, the idea that water is a malleable structure composed by a complex has been imposed as an undeniable principle, even without understanding the behavior of its organizational structure.

The dependence of life concerned with water can be noticed in all levels of human comprehension. The amount of water evaluated to be present in Earth is in the order of 1.39×10^9 km^3, which is equivalent to 1.39×10^{21} liters, being more than 97% composed by oceans. From the existent fresh-water (less than 3% of the total), more than 99% are unavailable for human consumption, being found in the glaciers of the polar regions and very profound subterranean water storages (Clarke and King, 2005).

It is not only the water dependence, expressed by means of the life conditions and the biological fundamental processes, which call for our attention to its fundamental contribution of several natural phenomena. Also, the specific characteristics of the water behavior, already well refined, bring to the belief that a better understanding is still essential.

Several characteristics of the water are extremely exotic, in the sense that they counteract the expectance of a physical-chemical characterization based on the study of similar molecules. The amounts of exotic characteristics of such a molecule, make water an extremely anomalous liquid (http://www.lsbu.ac.uk/water/anmlies.html).

Among the tens of anomalies presented by water, there are some characteristics we handle ordinarily without noting their importance. The fact that water is liquid in a wide range of temperature (between 0°C and 100°C) is surprising and absolutely

Universidade de São Paulo, São Paulo, Brazil. amiranda@if.usp.br

Presented at the XXth Symposium of GIRI, São Paulo, 2006

essential for the human life. Studies about the melting and ebullition points (Pashley et al., 2005, 2007; Francis et al., 2006), as a molecular mass function, show that molecules such as H_2S and H_2Se appear ordinarily as liquids under Earth pressure conditions in an 25°C interval only, and at temperatures below 0°C. Extrapolating the ebullition points of the 6A group hydrates found in the periodic table, the ebullition point of water should be about 150°C greater than expected.

Other extremely *exotic* but familiar characteristic is the fact that the water presents their maximum density at a temperature of about 4°C, remaining yet in the liquid state (at the normal environment pressure conditions). Lowering the temperature beyond their freezing point, the water presents a volume increase of approximately 9%. This water characteristic allows, among other many phenomena, the formation of the ice layer that fluctuates over the "frost sees", acting like a "thermal insulator" and helping to prevent the heat loss under the ice layer. This is essential for the sub-aquatic life maintenance.

Among several other anomalous characteristics of water, one can cite the high specific heath – 4.18 J g^{-1} K^{-1} (Wagner et al., 2000), high dielectric constant – 78.4 (Murrel and Jenkins, 1994), viscosity – 0.8909 mPa s (Cho et al., 1999), and surface tension – 0.07198 J m^{-2}, respectively at 25°C (Cooper, 1994).

One of the most intriguing anomalies presented by water is the so called (currently) *Mpemba effect*. It consists to observe that hot water reaches the freezing point faster than the colder one under certain circumstances. Although this effect had been previously observed by Aristotle, Roger Bacon, Francis Bacon and Descartes, among others, the scientific community has devoted little attention to this phenomenon until few years ago, when Erasto Mpemba, a scholar boy in Tanzania, has noticed this fact (Jeng, 2005, 2006; Auerbach, 1995).

Another characteristic, not less intriguing, is concerned to changes of super cooled water physical properties. Statistical models, based on molecule orientation variables, point to the coexistence of different densities even in a same pressure-temperature space region. This is directly related to the hydrogen bond dynamics (Choukroun and Grasset, 2007; Mishima and Stanley, 1998; Tomé, 2003).

Historically, despite some controversies found in literature, the discovery of water chemical composition is attributed to Henry Cavendish (1731–1810) who has observed experimentally, in 1781, that water is composed by hydrogen and oxygen (Seitz, 2005). Since 1972, models have been proposed to describe liquid water, supposing the formation of agglomerates of H_2O (clusters) in different sizes (Hagler and Scheraga, 1972). More recently, deeper studies about the electrical dipole moment and configurations of such heterogeneous clusters brought great contribution to understand the liquid structure of water (Gregory et al., 1997).

Recently, in addition to the numerous studies about water, the action of extremely diluted solutions has been considered a high interest research area (Rey, 1998; Lobyshev et al., 1999; Becchi et al., 2005; Barbero et al., 2007). The great majority of the proposed models is designed to explain the behavior of the ultra-high dilutions based on the capacity and complexity of water molecules organization (Rey, 1998; Lobyshev et al., 1999). An essential question that attempted to be elucidated is the way in which this dilutions affect the such organization and how specific it is.

Water Modeling

The water molecule possesses three vibration modes: η_1, η_2 and η_3, which represent, respectively, two stretching modes of the OH bonds, symmetrical and asymmetrical, and the angular deformation of the forming bonds molecules. When measurable quantities of molecular species dissolved in water are considered, the interactions can be identified by means of changes in the vibration modes, mainly in the modes η_1 and η_3.

The structural models of water may be divided in two main classes: mixture models and continuous models. The mixture models assume that discrete species, differing by specific structural arrangements, remain in equilibrium. Conversely, the continuous models assume that the water molecules are completely bonded by means of hydrogen bonds into a continuous lattice. In this case, the bonds distortions produce a continuous distribution of lengths, angles and energies (Rossi, 1995; Rossi et al., 1996; Libnau et al., 1994).

According to the version accepted by the researchers favorable to the mixture model, the mechanism of the equilibrium between two different species is used to explain the organizational structure of water. The molecules of water, in liquid state, are considered as forming units for the micro domains (clusters) that grow with decreasing temperature. The internal molecules of these clusters are considered as "bound water", and the external, "free water" molecules.

For simplicity, we can define the concepts of "free water" and "bound water" according to the relative amount of inter- and intra-molecular interactions (Fig. 4.1), i.e., if a certain set of water molecules shows many inter-molecular interactions, then, "bound water" molecules are present. Thus, "free water" molecules should not present many inter-molecular interactions. The presence of inter-molecular interactions must influence the specific characteristics of the liquid, such as the infrared absorption spectra, which should be sensitive to the related alterations.

Rossi (1995), looking for information about reaction kinetics, obtained very interesting results from water infrared absorption spectra, which have suggested the existence of a dynamic equilibrium between two "classes" of molecules. Working with HDO samples (in order to eliminate the coupling between the vibrations of

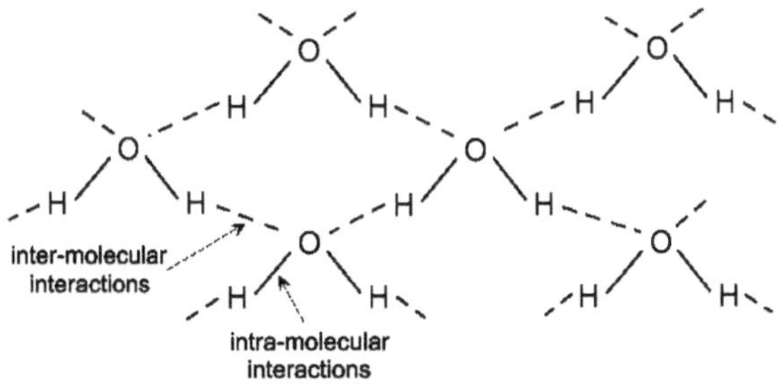

Fig. 4.1 Interactions sketch among water molecules

both modes, by introducing the mass difference of H and D); the author related spectral variations after changes of temperature. Two isosbestic points were observed in the spectrum for temperatures ranging from 5°C to 55°C, which were attributed to O-D (λ = 2,530 cm^{-1}) and O-H (λ = 3,450 cm^{-1}) stretching modes, respectively.

In an isosbestic point, different species present the same absorption. Thus, for a system containing two components in chemical equilibrium, this absorption does not change with their relative concentrations. The author have then concluded that the spectral variations of HDO according to temperature must be associated to the dynamical equilibrium between the two "classes" of molecules, it means, "free water" and "bound water".

The study of Rossi has called for attention because to the decrease of molecular interactions produced by increase of temperature. After the clusters model, it is possible to suppose that the decrease of intermolecular interactions modifies their physical structure. And, probably, the decrease of intermolecular interactions number in a solution could be related to the decrease of bigger clusters population and increase of smaller clusters. Of course, among other factors, this will be dependent of the different sized clusters stability and their intrinsic configurations. Moreover, a maximum limiting temperature should exist, above of what the formation of clusters should no more be allowed (Rossi et al., 1996; Libnau et al., 1994).

Study of Ultra-high Dilutions

The physical properties of water and aqueous solutions have been studied over many decades (Hagler and Scheraga, 1972). The studies found in the literature involve the monitoring of specific parameters, such as the electrical conductivity (Elia et al., 2000, 2004) emission and absorption of radiation (Rey, 1998, 2003; Lobyshev et al., 1999), refraction index and others. Important theoretical works (Errington and Debenedetti, 2001) have also demonstrated the need of understanding of the organizational behavior of molecules in aqueous solutions.

In contrast with other liquids, which in general have their behavior mainly related with Van der Waals interactions, the behavior of water is strongly related to the presence of hydrogen bounding. These interactions have approximately one order of magnitude more energy than those of the Van der Waals interactions, imposing a great ordering of the system, since a specific orientation between the water molecules is demanded.

The idea that water is composed by micro-domains has been strengthened during the last years (Gregory et al., 1997; Lobyshev, 1999; Rossi, 1995; Woutersen et al., 1997). Some works suggest that these micro-domains would have two types of structure and, perhaps, they could be dynamically converted one to another (Elia et al., 2004).

It is consensual in the literature that many peculiar characteristics of liquid water are related to hydrogen bonds. Although the "life time" of each hydrogen bond is very short and related to temperature (Lamanna et al., 1995), some researchers believe that the dynamics of water may be seen, during short intervals of time, as a structure composed by regions containing ordered and disordered clusters, similar to a defected crystal (Lobyshev et al., 1999).

Photo-Absorption and Photo-Emission

Lobyshev et al. (1999) have observed in water luminescence spectra that the emission spectrum of pure liquid water (at 20°C) contains two characteristic bands: one having its maximum at λ_{em}[1] = 360 nm ("short-wave" band), generated by the water molecules excited at λ_{exc}[2] = 260 nm, and the other of λ_{em} = 410 nm, excited at λ_{exc} = 310 nm ("long-wave" band).

As discussed above, different luminescence bands may be justified in two ways: by considering the presence of two types of chemically different substances, or by only one type of substance in two distinct ways. However, the above results are related with pure water samples, and such explanations cannot be accepted in a conventional way. Variation in experimentally observed luminescence intensity was, then, justified by the authors as changes in the efficiency of excitation energy transfer from the absorption to the emission centers (Lobyshev et al., 1999). Moreover, the authors proposed that this change of efficiency must originate in the organizational structure of water. Thus, changes in water properties must deeply influence the energy transfer efficiency.

After investigating the pure water intrinsic luminescence, the authors extended the study to the ultra-high dilutions of "luminescent" and "non-luminescent" substances. It was then found a non-monotonic dependence of the luminescence intensity *versus* the substance concentration, even if non-luminescent substances were used in solutions. These results strengthened the author's proposal that the change of the excitation energy transfer efficiency is related to aqueous solution organizational structure.

The hypothesis proposed is that, according to the classical concepts, for sufficient dilution so that the concentration value could be neglected (as well as for absorption effects), the luminescence intensity should depend linearly on the luminescent additive concentration and independent of the non-luminescent additives.

It was observed experimentally in 10^{-6}M solutions of glycylasparagine (non luminescent substance) that the "short wave" and "long wave" band intensities are two times bigger than in the case of pure water, without changing their peaks position in the spectrum. Also, the concentration reduction in new glycylasparagine solutions leads to a non-monotonic dependence of luminescence intensity.

[1] Emission wavelength.

[2] Excitation wavelength.

A non-monotonic dependence is also observed in solutions of glycyltryptophane (luminescent substance). However, it was observed a proportional "long wave" band increase in relation to concentration, which increased from 10^{-7}M to 10^{-4}M. The authors explain this behavior as a possible superposition of the natural glycyltryptophane luminescence band with that of water "long wave".

The basic proposal of the authors is that the energy is absorbed by all the "structure" of water and after, is transferred to the emission centers (structural "defects"). The primary reaction would not occur in the first excitation origin place, but only after its diffusion throughout the hydrogen bonds system, until some defect in the water structure is reached.

Supposing then that the excitation is absorbed by a "main lattice" and thereafter transmitted to the structural defects, the influence of a certain molecule would be defined by the dimension of the region that it is able to disturb, changing the balance between the order-disorder states of water.

Thermoluminescence

Considering that the thermoluminescence is an appropriate tool for the study of crystalline structures, ordered or disordered (because the luminescent centers appear in the regions containing lattice imperfections), L. Rey (2003) has investigated, by means of this technique, ultra-high dilutions of lithium chloride (LiCl) and of sodium chloride (NaCl).

Working with H_2O and D_2O at very low temperatures, the authors observed that the thermoluminescence spectra of frost samples, after irradiated with γ–rays, are consisted of two main peaks: peak 1 – near 120 K, and peak 2 – near 166 K. As expected, the D_2O samples showed a signal much more intense than H_2O. Moreover, it was well stated that the relative intensity of the thermoluminescent glow is determined by the applied irradiation dose.

One of the hypotheses proposed about the nature of the thermoluminescent emission of ice relates the peak 2 to the chain of hydrogen bonds; whereas peak 1 would be related to the individual characteristics of the water molecule. That way, it is expected that the peak 2 would be strongly related to the structure formed by the intermolecular bonds (such as the hydrogen bonds), whereas peak 1 should reflect, with more precision, some characteristics linked to the intra-molecular bonds.

One of the arguments used by Rey to corroborate this hypothesis is based on the results obtained from the luminescence of "formamide", well known to have strong hydrogen bonds. The thermoluminescence spectrum of this substance shows a well-pronounced peak in the corresponding region of the peak 2 of the ice spectrum, strengthening thus the hypothesis that this peak would be related to the hydrogen bonding chain.

Another interesting aspect discussed by Rey is referred to the results obtained with the thermoluminescence of LiCl solutions, known as efficient suppressors of hydrogen bonds. He verified experimentally that the peak 2 was totally extinguished, even at relatively low concentrations (0.1 M of LiCl), whereas the emission of peak 1 remained approximately unmodified (Rey, 2003).

Their most interesting results are concerned with thermoluminescence of ultra-high dilutions. Working with LiCl and NaCl, and using a centesimal dilution protocol accompanied with strong standard mechanical agitation, Rey has studied solutions with concentrations (theoretically calculated) of 10^{-30} g cm^{-3}, correspondent to the 15th dilution. After cooling the samples in order to reach sufficient stability in crystallization process and after x-ray irradiation at 77 K (with dose of 0.4 kGy), or with γ–ray (reaching a dose of 19 kGy), he obtained an emission spectra in the range of temperatures from 77 K to 213 K. After irradiation, the samples were stored in a recipient containing liquid nitrogen for one week. The thermoluminescent glow was compared among the three systems: (i) 15th dilution of LiCl in D_2O, (ii) 15th dilution of NaCl in D_2O, and (iii) pure D_2O submitted to the mechanical process of 15 dilutions. The three systems were found to be substantially different, both for the x-ray or γ-ray irradiations (Elia et al., 2000). By comparing these results it was possible to confirm the tendency of hydrogen bonds suppression in the LiCl solution, despite its ultra high dilution.

Further Work

Other works dealing with the organizational behavior of water in aqueous solutions have also been developed, outstanding the influence of the species involved in the solution, the dynamical orientational complexity of water molecules, and the importance of some characteristics such as size and shape of clusters (Gregory et al., 1997; Elia et al., 2000, 2004; Woutersen et al., 1997, 1998; Woutersen and Bakker, 1999; Mishima and Stanley, 1998; Rocha, 2001).

Samal and Geckeler (2001), for example, working with aqueous solutions of several substances (fullerene-cyclodextrin, β-cyclodextrin, sodium chloride, sodium guanosine monophosphate and a DNA oligonucleotide), observed increase in the natural aggregation of these substances when reducing their concentrations. They have distinguished also that the clusters formation mechanisms, although not yet well known, should depend on the nature of the involved species in the solution, as much of the solute as of the solvent.

Studying the vibrational and orientational dynamics of HDO dissolved in D_2O, Woutersen et al. (1997), on the other hand, have observed that the relaxation of the HDO molecules, regarding their orientation, occurs in a scale of time that either too fast or too slow (with time constants associated with $\tau_R = 13$ picoseconds and $\tau_R = 0.7$ picoseconds). The authors concluded that the water molecules with strong hydrogen bonds only relax by means of a slow process, whereas the fast process dominates for molecules of weak hydrogen bonds. The authors suggest the existence of two different "molecular species" present in the studied samples.

Gregory et al. (1997) carried out a theoretical and experimental work where the dipole moment was studied in water clusters of many sizes and configurations.

They concluded that in the condensed phase, where the average dipole moment of a water molecule is about 40% bigger than in the isolated monomer, can be well represented by a model that considers the presence of clusters of small size.

It is worth to consider also the calorimetric studies in ultra-high dilutions field (Elia et al., 2000). In these studies, measurements of electrical conductivity also have been indicated as a possible tool to enlighten the behavior of these solutions (Elia et al., 2004), but significant results have not yet been attained.

Present Research – Impedance Spectroscopy

For several years the impedance spectroscopy has been used as an important tool for the study of material behavior, as in the research centers, as in several sectors of industry involving new materials manufacture, quality control of pharmaceutics products, electronic component parts, production of glasses, technology of thin films, corrosion studies of several kinds of materials and many others.

By impedance spectroscopy, it is possible to study the dielectric behavior of several materials, in order to analyze the electrical responses of solid and liquid electrolytes and characterize some electrical components, such as capacitors, resistors and inductors. It allows access informations about very complex systems that may be compared under the electrochemical point of view (Macdonald, 1987).

The electrical impedance (Z), a more general concept than resistance (R), was introduced for the first time by Oliver Heaviside in 1880 and was developed through a vectorial modeling in a complex plane by A. E. Kennely and C. P. Seinmetz (Macdonald and Ross, 1987).

It is important to observe that the complex electrical impedance is, by definition, time independent variable (despite of the applied tension and the resultant current being time dependent), but it is dependent of the angular frequency (ω) of the applied tension.

Thus, in order to study the dielectric properties by means of the impedance spectroscopy, the material to be analyzed must be introduced between the plates of a capacitor. The behavior of the real (Z') and the imaginary part (Z'') of the system impedance are analyzed, in a chosen interval of frequencies.

The impedance analysis consists in the study of Z behavior as a function of ω in a frequency interval that provides information about the properties of the medium. This analysis is done by considering that there is a relation between the electrochemistry of the medium and an idealized equivalent circuit, composed of elements such as equivalent resistors and capacitors (represented by their respective resistances R and capacitances C).

The conclusions obtained from the analysis of these behavior (of an idealized circuit) would be extended to the dynamics of the studied material. But it is necessary to guarantee that the fundamental laws related to the charges and potentials and the linear properties of the system are not violated.

The diagram shown in Fig. 4.2 (graphical analysis of the Z' and Z'' values in the complex plane, known as the Nyquist diagram) represents the typical response

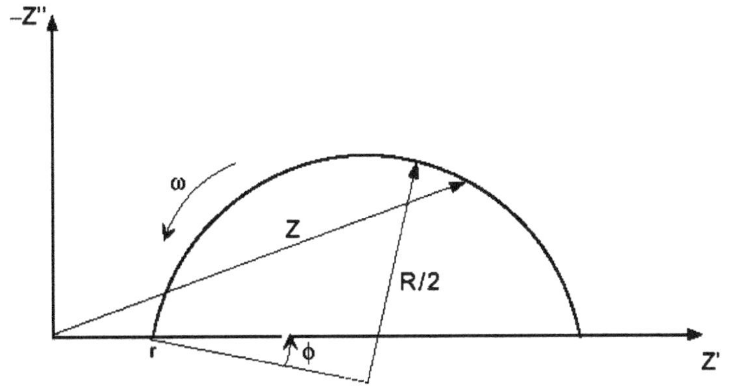

Fig. 4.2 Impedance diagram in the complex plane, characteristic of a circuit composed by a resistor (r) in series with a parallel circuit RC, where ø is the angle of decentralization of the semicircle formed by the values of –Z″ as a function of Z′, when it occurs a variation of ω

obtained for liquid and solid materials when submitted to impedance analysis. Another form, frequently used to study the impedance of the system, consists in the behavior analysis of both components of Z as a function of ω, separately (Z′ × ω and Z″ × ω).

Some materials have constituents that present distinct contributions to their electrochemical behavior. They can be often observed by specific characteristics of the impedance curves at different frequency regions. Each material constituent changes the impedance curve in a well-defined frequency range, different for each constituent. In such cases, it is possible to compare the system with circuits composed by the association of RC parallel sets, linked together. In this case, the corresponding impedance diagram (Fig. 4.2) is composed by more than one semicircle, each one being referred to a specific contribution of the material. When these contributions are not perfectly distinguished in regard to the excitation specific frequencies, the system equivalent circuit can be much more complex. In this situation, the semicircles frequently appear superposed, making more difficult the data analysis.

As it was pointed before, it is essential that the fundamental laws that rule the materials behavior should not be violated by the mathematical modeling applied in the data analysis. In the case of water and diluted solutions, some works of impedance spectroscopy were already carried out. However, only solutions with concentrations higher than 10^{-4}M were investigated since the main objective was to study the dynamics of the ions diffusion (Becchi et al., 2005). The literature of impedance spectroscopy of pure water is concentrated in the range of frequencies lower than 1 kHz (Grasso et al., 1990).

Studies dealing with the dielectric constant of water are currently performed at frequencies in the order of MHz, assuming that both, the dielectric constant and the characteristic frequency are independent of ω in this range.

According to the classical model proposed by Gouy, Chapman and Stern (GCS), the liquid-electrode interface may be described by the two-layer composition: the

"Helmholtz layer" – more external and related with the absorption of ions at the electrode surface; and the "diffusion layer" – related with charge diffusion between the representative volume of the material (bulk) and the Helmholtz layer.

Apparently, in the low frequency range (<1 kHz), the interface effects, that could disturb the characteristics of the solution dielectric behavior, should not be neglected when pure water and extremely diluted solutions (concentrations lower than 10 ppm) are studied. However, in the range of higher frequencies (greater than 1 kHz), it is expected that the diffusion velocity of the ions could be neglected, minimizing the possibility of the occurrence of interface effects.

Grasso et al. (1990), in a study on the influence of the diffusion layer in water impedance measurements, showed that the capacitance and the resistance values strongly depend on the frequency (for the frequency ranges lower than 1 kHz and 1 Hz, respectively). They attributed this phenomenon to the influence of the interface effects. However, in the interval between 10^3 Hz and 10^5 Hz, both the capacitance and the resistance of the sample cell, fulfilled with water, remained constant.

Another extremely important aspect for the application of the impedance spectroscopy technique is related to the fact that non-linear electrochemical are the most representative of the real electrode-material systems. For the non-linear systems, the impedance measurements are significant (in the description of the studied material) only when the signal magnitude shows that the global response of the electrode-material system is electrically linear. Moreover, the non-linear behavior of the electrochemical systems may become very important when high voltages (or currents) are applied; and frequently is related with the influence of the interfacial defects. On the other hand, even systems that show a strongly non-linear response (when submitted to high voltages) can exhibit a linear behavior under low applied tension (Macdonald and Ross, 1987).

The ion diffusion dynamics has been investigated in salt solutions with concentrations down to 10^{-4}M (Barbero et al., 2007). However, in a frequency range where the interface effects can be minimized, the study of water and ultra-high dilutions (concentrations lower than 10^{-4}M) constitutes a new question not yet well understood.

Impedance Analysis Applied to the Study of Pure Water and Ultra-High Dilutions

Based on the above summarized revision about ultra-high dilutions research, carried out during the last years, and also on the use of the impedance spectroscopy as a tool for the study of aqueous solutions and of pure water, we have developed this work, as explained below (Miranda, 2008).

The purpose of this work was to study the impedance of pure water and lithium chloride solutions (substance recognized as a suppressor of hydrogen bonds), in a frequency range between 1 kHz and 13 MHz.

The method used to prepare the samples was the Hahnemannian Centesimal Method described in the Brazilian Homeopathic Pharmacopoeia (1997) and in the

Technical Standards Manual (2003). The dilution was carried out with distilled water for all samples. A set of 15 samples of lithium chloride solutions was prepared with concentrations ranging between 2.4×10^{-1}M and 2.4×10^{-29}M.

The equipment used to distil the water was a commercial distiller model DG-4K Gehaka, incorporated to a filter, both maintained under rigorous conditions of periodic maintenance. The used water was systematically submitted to analysis by the sanitary control, and the quality of the distilled water was monthly controlled by analyzing its physical-chemical parameters.

The succussion process, performed according to the Hahnemanian Centesimal Method, was done with a mechanical shaking device for all samples. This device can handle four recipients at a time, which promotes, by means of a "mechanical arm", successive impacts on a semi-rigid screen with constant frequency and intensity.

To prepare the samples, the dilution percentage were calculated in weight and, in all stages of weighing, a precision digital balance Gehaka BG4400 was used, which is calibrated by the technical team and also, at each six months, by the Brazilian Calibration Center Ltd. (Laboratory of Metrology – CEBRAC). The whole process of sample preparation was carried out in a room equipped with temperature and humidity control.

The lithium chloride (LiCl) matrix was purchased from Sigma-Aldrich and their minimum purity is 99.0%. To avoid any possible influence of the sample preparation process in the measurements – which could artificially modify the conductivity (or any other physical characteristic) of the analyzed solutions – a special care was exercised. Thus, for comparative analysis, pure water samples submitted to identical process were analyzed in parallel of the saline samples, from the 1st to the 15th dilution and succussion stage. This way, each set of analyzed LiCl samples was accompanied with a set of "diluted" pure and succussed water (always with 15 corresponding samples). It is important to emphasize that the water used for the production of a given set of samples was obtained from the same distillation, including the set of saline samples and the set of pure water samples.

The equipment used for the experimental measurements was an Impedance Analyzer, HP model 4192A, that possesses an internal frequency synthesizer capable to generate signals between 5 Hz and 13 MHz, with resolution of 1 mHz. The sample cell was constructed so that an insulator material, being the sample confined to this restricted internal place, wrapped the capacitor. Thus, through the real and imaginary part of the impedance, the resistive and reactive characteristics of a circuit, which represents the behavior of the studied samples submitted to a sinusoidal tension of variable frequency, were studied. The characteristic parameters of the equivalent circuit were obtained with the aid of the Z-View software.

In order to avoid possible contamination (such as the development of fungus, bacteria, etc.) it was established that the samples would be utilized only for a period of ten days after their preparation. During this period the samples were stored at 10°C. The fraction of the sample fulfilling the measurement cell was discarded after use.

The measurements were done at various temperatures: near room temperature, closed to 0°C and −70°C, and at liquid nitrogen temperature. However, in the frequency range of 1kHz to 13MHz, the most interesting results are related with the measurements performed at room temperature. Measurements performed with crystallized samples might provide relevant information about pure water and ultra-high dilutions, but the adopted measurement procedure should be adapted in order that low temperature measurements could be done, fulfilling the required reproducibility conditions. We intend to do that in the near future. For all the measurements the temperature was carefully controlled in order to assure a good reproducibility of the results under the applied experimental conditions.

The experimental results of the ultra-high dilutions of LiCl showed that the imaginary part of impedance, for concentrations between 10^{-6}M and 10^{-29}M, presented a similar behavior of the Z″ measured for the pure water. A minimum was observed in the experimental curves obtained from the imaginary part of the impedance as a function of frequency, characterizing a "peak" between 10^4 Hz and 10^6 Hz. This resonant behavior occurs at a frequency for which the interface effects (related to the intermediary region between the bulk and the capacitor plates) could be neglected (Macdonald and Ross, 1987).

As the 1st and the 2nd dilutions of LiCl contain reasonably high concentrations, it is expected that their behavior be dominated by the presence of the salt. The experimental results show that the impedance parameters of these solutions are completely different from those of pure water and of the remaining ultra-high dilutions. Therefore, the study of the 1st and the 2nd dilutions of LiCl did not take part of the present study.

About the comparison between the LiCl solutions and pure water, it was possible to observe significant differences in peak intensity and frequency, set between 3rd and 15th. The 3rd, 7th and the 12th LiCl dilutions presented the greater differences in relation to pure water. An example of this difference can be observed in Fig. 4.3, where the curves of the imaginary part of the impedance (Z″) as a function of frequency (f) are shown, for the 12th LiCl dilution and pure water, respectively.

Concerning pure water, the experimental data showed to be highly promising for more accurate studies of impedance spectroscopy. The parameters measured in this work already indicate the influence of specific characteristics of water when submitted to a dilution and succussion processes, having received or not some initial molecular information.

The more adequate hypothesis that was considered is that there would be an effective change in the structure of water, caused by the initial presence of the solute and then diffused during the dilution process, with the help of the energy inserted by succussion.

If there are indeed changes in water structure, these might or not be related to different sizes and shapes clusters formation. The results of the equivalent circuit analysis have shown that the impedance data obtained from ultra-high dilutions and pure water are well modeled by a circuit composed by a series resistance r followed by a parallel RC association. The resistance r was approximately constant for all the measurements, much smaller than R.

It was observed experimentally that the equivalent resistance R is an effectively representative parameter of the modeled data. This result is coherent with the information found in the literature concerning this type of analysis.

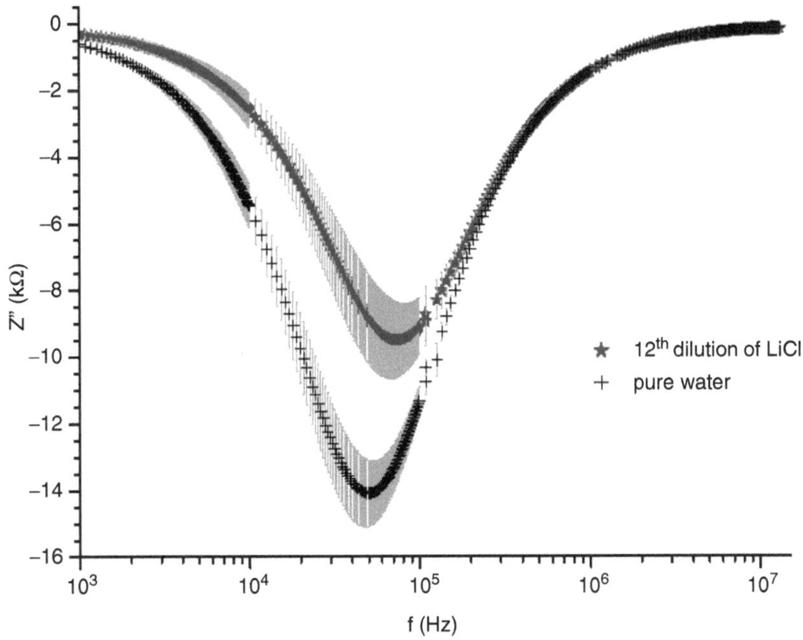

Fig. 4.3 Average curves from nine different measurements of the imaginary part of the impedance (Z″) as a function of frequency (f) for the 12th LiCl dilution (10^{-24}M) and of pure water, in the frequency range of 1 kHz and 13 MHz

This way, one contribution of this study was to show that the imaginary part of the impedance measured in concentrations much lower than 10^{-6}M present a behavior similar to that of Z″ that was measured in pure water. Moreover, the observation of the significant differences in frequency and intensity of the Z″ between LiCl solutions (dilutions between the 3rd and 15th) and pure water suggests that the technique of impedance analysis is sensitive to alterations in LiCl high dilutions properties. Yet, the experimental data suggest the possibility to differentiate patterns of different LiCl dilutions, even when concentrations are lower than 10^{-6}M.

The found results strongly suggest that testing solutes specificity through the observation of high dilutions - whose concentration is sufficiently small that their direct molecular influence no more could be considered – could be able to bring important information about structural changes in the water.

Acknowledgments We thank to Prof. Walter Maigon Pontuschka by the use of the Impedance Analyzer and by a important help with the manuscript; to Cilene and Ademir to make possible the use of samples produced in a rigorous way, according to the necessary requirements stated by the objectives of this study; and to Prof. Valmir Antonio Chitta for the valuable scientific discussions.

References

Auerbach D. Supercooling and the Mpemba effect – when hot water freezes quicker than cold. Am. J. Phys.; 1995, 63 (10): 882–885.

Barbero G, Becchi M, Strigazzi A, Le Digabel J, Figueiredo Neto AM. Experimental evidence for the adsorption-desorption phenomenon on the spectroscopy impedance measurements of an electrolytic cell. J. Appl. Phys.; 2007, 101 (4): Art. No. 044102.

Becchi M, Avendaño C, Barbero G. Impedance spectroscopy of water solutions: The role of ions at the liquid-electrode interface. J. Phys. Chem. B; 2005, 109 (49): 23444–23449.

Brazilian Homeopathic Pharmacopoeia – Part I General Methods (in Portuguese), 2nd ed. São Paulo: Atheneu editora São Paulo Ltda.; 1997.

Cho CH, Urquidi J, Singh S, Robinson GW. Thermal offset viscosities of liquid H_2O, D_2O, and T_2O. J. Phys. Chem. B; 1999, 103: 1991–1994.

Choukroun M, Grasset O. Thermodynamic model for water and high-pressure ices up to 2.2 GPa and down to the metastable domain. J. Chem. Phys.; 2007, 127 (12): Art. No. 124506.

Clarke R, King J. The Atlas of the Water (in Portuguese). São Paulo: Publifolha; 2005.

Cooper JR. IAPWS Release on Surface Tension of Ordinary Water Substance. International Association for the Properties of Water and Steam; 1994.

Elia V, Niccoli M. New phisico-chemical properties of water induced by mechanical treatments. A calorimetric study at 25°C. J. Therm. Anal. Calorimet.; 2000, 61: 527–537.

Elia V, Napoli E, Niccoli M, Nonatelli L, Ramaglia A, Ventimiglia E. New physico-chemical properties of extremely diluted aqueous solutions. A calorimetric and conductivity study at 25°C. J. Therm. Anal. Calorimet.; 2004, 75: 815–836.

Errington JR, Debenedetti PG. Relationship between structural order and the anomalies of liquid water. Nature; 2001, 409: 318–321.

Francis MJ, Gulati N, Pashley RM. The dispersion of natural oils in de-gassed water. J. Coll. Interface Sci.; 2006, 299: 673–677.

Grasso F, Musumeci F, Triglia A. Impedance spectroscopy of pure water in the 0.01 Hz to 100 kHz range. Nuovo Cimento; 1990, 12D(8): 1117–1130.

Gregory JK, Clary DC, Liu K, Brown MG, Saykally RJ. The water dipole moment in water clusters. Science; 1997, 275: 814–817.

Hagler AT, Scheraga HA. Structure of liquid water. Statistical thermodynamic theory. J. Phys. Chem.; 1972, 76 (22): 3229–3243.

Jeng M. Hot water can freeze faster than cold?!?. arXiv: Physics/0512262; 2005

Jeng M. The Mpemba effect: When can hot water freeze faster than cold? Am. J. Phys.; 2006, 74 (6): 514–522.

Lamanna R, Floridi G, Cannistraro S. Cooperativity and hydrogen bond network lifetime in liquid water. Phys. Rev. E; 1995, 52: 4529–4532.

Libnau FO, Toft J, Christy AA, Kvalheim OM. Structure of liquid water determined from infrared temperature profiling and evolutionary curve resolution. J. Am. Chem. Soc.; 1994, 116: 8311–8316.

Lobyshev VI, Shikhlinskaya RE, Ryzhikov BD. Experimental evidence for intrinsic luminescence of water. J. Mol. Liq.; 1999, 82: 73–81.

Macdonald JR. Impedance Spectroscopy – Emphasizing Solid Materials and Systems. New York: John Wiley & Sons; 1987.

Miranda AR. Comparative studies of ultra-high dilutions of LiCl: Impedance spectroscopy in the frequency range of 1 kHz to 13 MHz (in Portuguese). Ph.D. Thesis. São Paulo: Physics Institute, University of São Paulo, Brazil; 2008.

Mishima O, Stanley E. The relationship between liquid, supercooled and glassy water. Nature; 1998, 396: 329–335.

Murrel JN, Jenkins AD. Properties of Liquids and Solutions, 2nd Ed. Chichester: John Wiley & Sons; 1994.

Pashley RM, Rzechowicz M, Pashley LR, Francis MJ. De-gassed water is a better cleaning agent. J. Phys. Chem. B; 2005, 109: 1231–1238.
Pashley RM, Francis MJ, Rzechowicz M. The hydrophobicity of non-aqueous liquids and their dispersion in water under de-gassed conditions. Curr. Opin. Coll. Interface Sci.; 2007, doi: 10.1016/j.cocis.2007.07.005.
Rey L. Low temperature thermoluminescence. Nature; 1998, 391: 418.
Rey L. Thermoluminescence of ultra-high dilutions of lithium chloride and sodium chloride. Physica A; 2003, 323: 67–74.
Rocha MSF. Process of dieletric relaxation in glass BABAL irradiate with x-rays (in Portuguese). Ph.D. Thesis. São Paulo: Physics Institute, University of São Paulo, Brazil; 2001.
Rossi AV. Kinetics studies of the reactions of PAR with Zn^{2+} e Cu^{2+} and spectroscopy of ATR studies of saline aqueous solutions: Investigation about the influence of the organization of aqueous solutions in chemical reactions (in Portuguese). Ph.D. Thesis. Campinas: Chemistry Institute, University of Campinas, Brazil; 1995.
Rossi AV, Davanzo CU, Tubino M. The structure of liquid water in aqueous solutions: A proposed model from near and mid infrared spectroscopy. J. Brazilian Chem. Soc.; 1996, 7 (6): 403–410.
Samal S, Geckeler KE. Unexpected solute aggregation in water on dilution. Chem. Commun.; 2001, 21: 2224–2225.
Seitz F. Henry Cavendish: The catalyst for the chemical revolution. Notes Rec. R. Soc.; 2005, 59: 175–199.
Technical Standards Manual for Pharmacy Homeopatic (in Portuguese), 3rd ed. Ampliação dos Aspectos Técnicos e Práticos das Preparações Homeopáticas, ABFH. São Paulo: Associação Brasileira de Farmacêuticos Homeopatas; 2003.
Tomé T. (Ed.) Tendencies of the statistical physics in Brazil (in Portuguese). São Paulo: Editora Livraria da Física; 2003.
Wagner W, Cooper JR, Dittmann A, Kijima J, Kretzschmar HJ, Kruse A, Mare R, Oguchi K, Sato H, Stöcker I, Sifner O, Takaishi Y, Tanishita I, Trübenbach J, Willkommen TH. The IAPWS industrial formulation 1997 for the thermodynamic properties of water and steam. ASME J. Eng. Gas Turbines Power; 2000, 122: 150–182.
Woutersen S, Bakker HJ. Resonant intermolecular transfer of vibrational energy in liquid water. Nature; 1999, 402: 507–509.
Woutersen S, Emmerichs U, Bakker HJ. Femtosecond mid-IR pump-probe spectroscopy of liquid water: Evidence for a two-component structure. Science; 1997, 278: 658–660.
Woutersen S, Emmerichs U, Nienhuys HK, Bakker HJ. Anomalous temperature dependence of vibrational lifetimes in water and ice. Phys. Rev. Lett.; 1998, 81 (5): 1106–1109.

Part III
Basic Research: Biology

Chapter 5
Non-linearity Modeling of Ultra-dilutions: The Histamine Disturbances Case

Guadalupe Ruiz-Vega and Gabino Estevez-Delgado

Introduction

The knowledge about the world and the laws that govern natural phenomena has been one of mankind's most persistent interests; theoretic and empiric models are the usual tools to achieve this goal.

In natural sciences, a model is a description of how something works. Descriptive tools are diverse, depending on the characteristics of the system and the model selected (for example: biological, physical, or mathematical models). Modeling is a recursive process, in most cases also iterative, that confronts experimental data (meaning the real world) against theory (Montero and Moran, 1992). The theoretical analysis of an experimental system is done through a gradual abstraction, in order to grasp the features that characterize the system's behavior. In this stage, the hypotheses, assumptions, approximations, and theories on the subject are taken into account in order to structure the model.

Every model is comprised of a set of parameters, either constant (also called model parameters) or variable, which function as inputs (e.g. dose) or outputs (e.g. response). In addition, parameter evaluation techniques can range from simple, such as a linear regression, to extremely complicated, such as non-linear approaches; the best fit should be the one that resembles the system's behavior in the real world.

In general, mathematical models of dose-response can be:

a. *Empirical models*
The goal of empirical dose-response modeling is to find a simple mathematical model to satisfactorily describe this pattern. Examples of empirical models include linear functions (such as those used in linear regression), log-linear models, Poisson regression (commonly used in epidemiology), and Hill models (commonly used to analyze ligand-receptor data).

Universidad Michoacana de San Nicolás de Hidalgo, Michoacán, México.
guadalupe.ruizvega@gmail.com

Presented part at the XVIII Symposium of GIRI, Brussels, 2004 and part at the XX Symposium of GIRI, São Paulo, 2006

L.V. Bonamin (ed.) *Signals and Images*,

b. *Models based on action mechanism*
In contrast to empirical modeling, mechanism-based modeling tries to grasp the mechanistic relationship between exposure (dose) and possible thresholds in order to describe the observed response simultaneously.

No matter the approach, in both cases it is necessary to identify the dose as well as the associated or expected response.

Dose-Response in Ultra-Low-Dose, ULD, Compounds

The Problem of Dose

It is critical when performing dose-response analyses to understand what is meant by "dose" and how it applies to the response. In dose-response modeling, "dose" is a broad term that would include the exposure to diverse stimuli, i.e. mechanical, pharmacological, electromagnetic, etc. In the pharmacological area, the term "dose" includes the amount of drug given to an experimental subject (animal, volunteer), by some specific route, at some specific frequency. In general, "dose" units should reflect the magnitude of the exposure and the frequency over which it applies. The "dose" can be expressed in a multitude of metrics, as molarities, ppm, mg/body weight, etc.

In contrast to pharmacokinetic laws and their conventional terminology meaning *median lethal dose,* $LD_{50,}$ *median effective dose,* $ED_{50,}$ *and the therapeutic index* (Goodman and Gilman, 1996), the term *dose* in the realm of ULD therapy is not completely understood. Such terms design macroscopic quantities; meanwhile the molecular presence of the active compound in ULD ($\geq 10^{-24}$ M) is practically zero. Hence it is not possible to use those expressions and the main rules around them, as the well known receptor-ligand theory.

According to experimental findings, the therapeutic efficacy of ULD compounds rests on the consecutive dilution/shaking steps. As the steps number increases, the concentration diminishes giving as a result an extremely diluted compound for the high potencies. From this point of view, in this field, the close equivalent to *dose* is the number of dilution steps.

The classical concentration-effect and dose-effect curves do not match with the empirical criteria (Vannier, 1991) in ULD practice which associates a deeper effect as the dilution increases; even more, an oscillatory pattern has been reported. Issues as frequency or half-life of doses are also unresolved.

Response

In this framework, "response" generally refers to an "event" or "change" perceived in an animal or a human following exposure to the stimulus. These responses can range from early (like biochemical alterations or homeostatic reactions), to more complicated ones (such as the development of chronic defects).

Operationally, it is important to know that dose-response relationship exists in two very different forms:

a. *Continuous dose response:* is the dose-response curve based on the intensity of the response, also known as non-stochastic response; it is essentially a matter of how large an effect is brought about in an individual.

b. *Stochastic response:* The second curve is based on the fraction of a population that responds to a given dose, in an "all or nothing" manner: in health risk assessment the focus is generally on how many subjects are affected rather than how severe the effect is.

Method

We attempt to assess the biological effect, meaning the response, of extremely diluted compounds by means of physical-chemistry parameters. These parameters should be objective, systematic and repetitive under methodical procedures. Our method relies on the fact that living beings emit electric signals. According to their source, these signals could be of different nature, and so their detection, e.g. electroencephalogram, EEG, electrocardiogram, ECG, Pletismometry, etc.

It is expected that, when a stimulus is applied, those physiologic voltages would change, perhaps in a subtle way. Mathematical analysis (linear and no-linear) would reveal those slight changes, which must be systematic, repetitive and objectively related to the expected physiologic outcome. We have focused on EEG (Ruiz and Torres, 1997; Ruiz-Vega et al., 2000; Ruiz-Vega et al., 2002; Ruiz-Vega et al., 2005) and ECG (Ruiz et al., 1999) case studies, as their alterations are related to relevant health disturbances. Particularly, the record of these signals through sleeping time and their further analysis, would avoid bias due to possible interference of subject zdiluted histamine and sleep.

Homeostasis and Histamine

Homeostasis is one of the most remarkable and most typical properties of highly complex open systems. A homeostatic system (an industrial firm, a large organization, a cell) is an open system that maintains its structure and functions by means of a multiplicity of dynamic equilibriums rigorously controlled by interdependent regulation mechanisms.

Such a system reacts to every change in the environment, or to every random disturbance, through a series of modifications of equal size and opposite direction to those that created the disturbance. The goal of these modifications is to maintain the internal balance.

From a dynamic point of view, when a system gets away from equilibrium non-linearity appears, due to mechanisms that generate instability. Kinetic analysis allows the assessment of the rate of change of the system's properties; in fact, this has been the most useful tool to study the response of a living being under a stimulus.

Sleep loss, either pharmacological or mechanical, induces a balancing counter-action, called "sleep homeostasis" (Schwierin et al., 1996), which evolves as an increase in non-REM sleep intensity. The main marker of the intensity in non-REM sleep is EEG slow-wave activity, defined as **the spectral density in delta band.**

After the discovery of how histamine is both a neurotransmitter and/or a neuro-modulator, several studies have explored the role of the brain histaminergic system in sleep-wake control (Monti, 1993). Current data suggests that this system plays an important role in wakefulness. On this regard, although the immuno-modulator effect of highly diluted histamine in basophile activation has been widely tested (Poitevin, 1992; Sainte-Laudy and Belon, 1993), its role as a neurotransmitter has not been studied yet.

In this context, we attempt to find an empirical dose-response model for ultra-diluted histamine, analyzing its effect on the awakening mechanism by measuring changes on the sleep pattern of Wistar rats.

Impairing or blocking the transmission of histamine increases the slow cortical activity and slow wave sleep, whereas enhancements of histaminergic activity promote wakefulness. As the ultra diluted histamine was applied at the beginning of the sleep cycle of healthy subjects, according to the *iso-endopathy principle,* it would induce an increase in wakefulness, which in turn would promote a homeostatic host response. To evaluate it, the EEG was recorded on parietal region during rats' sleep cycle and the response was evaluated by means of the spectral density in delta (0.5–2.5 Hz) band. This marker, already tested in former papers (Ruiz-Vega et al., 2003; Ruiz-Vega et al., 2005), was analyzed vs. time in seven randomized verum/control groups. Dosing for verum ranges from approximately 10^{-12}, to 10^{-60} M. Complete details of the method can be found in Ruiz-Vega et al. (2005).

Animals

Young adult male Wistar rats weighting between 250–350 g were used. All procedures complied with our Federal Regulations for the Use and Care of Laboratory Animals (NOM-062-ZOO-1999, Mexico). Rats were feed *ad libitum* on standard diet (Purina, Mexico).

Rats were anaesthetized (40 mg/kg pentobarbital), three 1.5 mm diameter stainless steel electrodes (resistance $< 0.1\ \Omega$) were implanted in the animals' cranium after trepanation with a hand-held Foredom drill, secured with dental resin and connected to a 15A54 Grass multichannel amplifier (Astro-Med, Inc. West Warwick, RI, USA). A bipolar EEG was recorded through two of these electrodes, which were implanted bilaterally in the parietal region, and the third one in the frontal area for reference (ground). Trepanation points were spotted stereotaxically at the following coordinates: 3.8 mm posterior to Bregma and 4.5 mm lateral to the side of central line for the bilateral electrodes (Paxinos and Watson, 1986). After surgery the animals were allowed to recover for six days; a further one-day habituation period took place, in which the electrodes were connected to the signal amplifier by a cable that allowed movement freedom. During the pre-operation and recovery periods, rats were maintained under a steady light-dark cycle. The next day measurements for treatment started.

Chemicals

Ultra diluted histamine and the solvent (87% alcohol/water) were acquired from Laboratorios Medicor®, Schwabe Mexican Group. Preparation for each of the high dilutions tested involved 6, 9, 12, 15, 18, 21, 24, 30 successive centesimal shake/dilution (1/99% vol.) steps, which approximately correspond to 10^{-12}, 10^{-18}, 10^{-24}, 10^{-30}, 10^{-36}, 10^{-42}, 10^{-48},10^{-60} M.

Experimental Set Up

In this experiment, the rats were housed individually in wooden cages built to avoid external disturbances while at the same time allowing observation. Day-lighting was provided so as not to alter circadian histamine pattern (Tuomisto, 1991). Each cage also has Faraday insulation to prevent interferences.

Each subject was connected to a 15A54 Grass amplifier module, being its controls set as follows: sensitivity 20.0 μV/div, display gain 1, band-pass filter 0.3–30 Hz, line filter on. The output was sampled at 8 Hz and online digitized with a National Instruments AT-MIO-64E-3 card and PolyView® v2 software (Astro-Med). Output was stored in a computer for mathematical analysis. Sampling frequency was chosen taking into account delta band features (Rechtschaffen and Kales, 1968) as well as the Nyquist theorem.

Experimental Procedure

Randomized groups of 12 animals for verum and 12 for control were formed for each dilution to test. Six rats were tested at a time, one to a cage. At this time rats were fed their regular diet and purified water *ad libitum*. At 10h00, $t = 0$ (t, time in hours) the first oral dose was provided to verum and control groups. Ultra-diluted histamine was provided orally (0.05 ml every 20 minutes) to treated groups during the first 2 hours; solvent was administered to control in similar circumstances.

Stimuli were provided from identical bottles, which were identified with a code. This code was stored in a closed envelope, which was opened until the mathematical analysis of all files was finished.

All records were gathered from T_0 throughout an 8-hour sleep period of the animals. On the completion of the recording session, the tests concluded for the six subjects and the process was repeated with six new ones until the whole 12/12 series was tested.

Mathematical Analysis

The recorded files were saved as digitized time series for further analysis. Each record was split into 1-hour periods and before carrying out any mathematical operation, every file was visually examined and noisy segments stemming from occasional disconnections or external artifacts (body movement) were eliminated and omitted from further analysis. In fact, due to excessive noisy events and diverse technical problems (a loose electrode, electric power off) some files in control and verum groups were discarded or even not recorded, final number of registers are listed in legends of Figs. 5.1a and b.

Mathematical analysis of data was carried out by somebody who did not (and could not) know the specific source of the file under study 'blind'. All series of 1

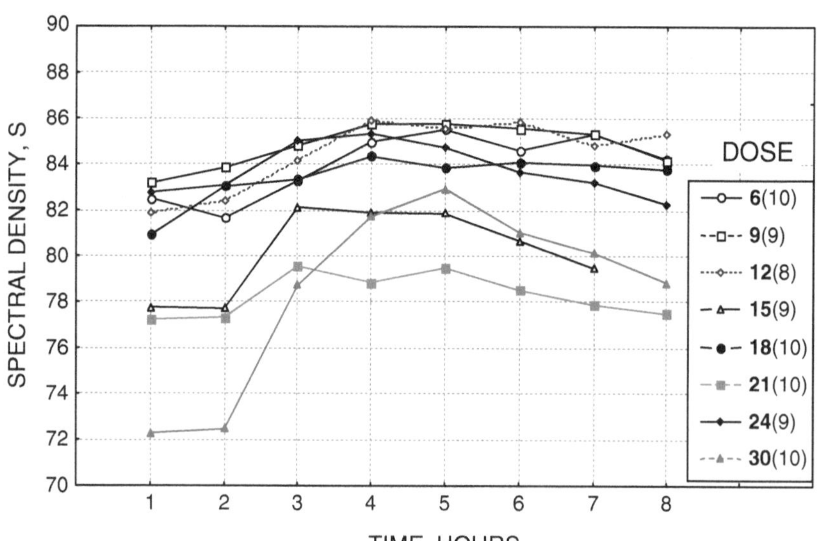

Fig. 5.1a Average values of spectral density in delta band *vs*. time for ULD-histamine groups. *Legend:* Ultra-Low-Dose as the number of centesimal shake/dilution steps in bold; number of animals in brackets

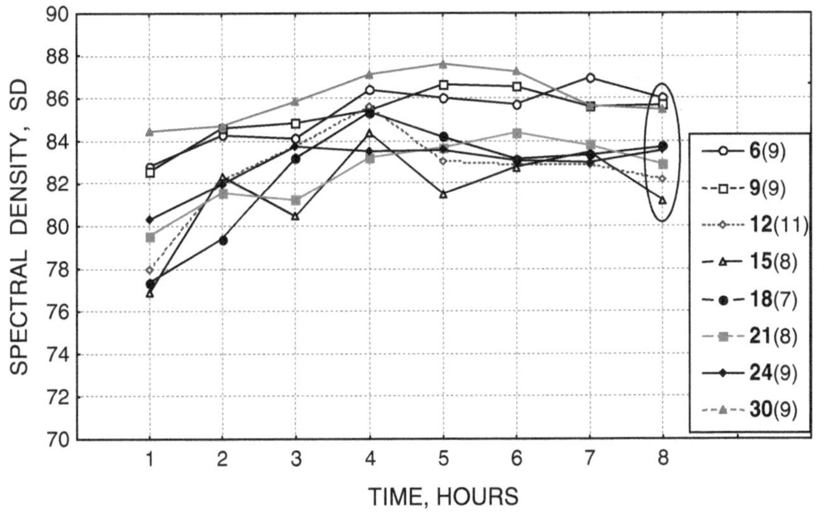

Fig. 5.1b Average values of spectral density in delta band *vs.* time for controls of ULD groups in Fig. 5.1a. Time (hours) in the x axis. *Legend:* ULD-histamine group in bold; number of animals in brackets. In all groups it is possible to observe an immediate increase in the marker with no statistical difference among them, running to an equifinal state, circled

hour data were subjected to a Fast Fourier Transform (FFT) algorithm and the spectral density in delta (0.5–2.5 Hz) band was computed for the specified bands in every file (STATISTICA®, StatSoft, Tulsa, OK, USA). Due to different lengths of noisy periods, the records were of different sizes. In order to have a common reference, the spectral density in the delta band was analyzed as percentage of the spectral density in the whole frequency range (0–2.5 Hz).

Comparison between-groups was conducted through an ANOVA and Shapiro-Wilk test for checking normality. We also performed the Newman-Keuls *post hoc* test. To avoid bias relating to extreme values, all outliers were identified and excluded (an outlier is an observation that is outside the range of ± 2 standard deviations). The level of significance was set at $p < 0.05$.

Results

The system's response can be analyzed from two approaches:

a. *The absolute value of the marker*

Results are shown in Figs. 5.1a, b, 5.2a, and b.

A gradual reduction in spectral density appears during dosage (the first 2 hours); but the changes are not consecutive, that is: values in group 15 diminish, but not

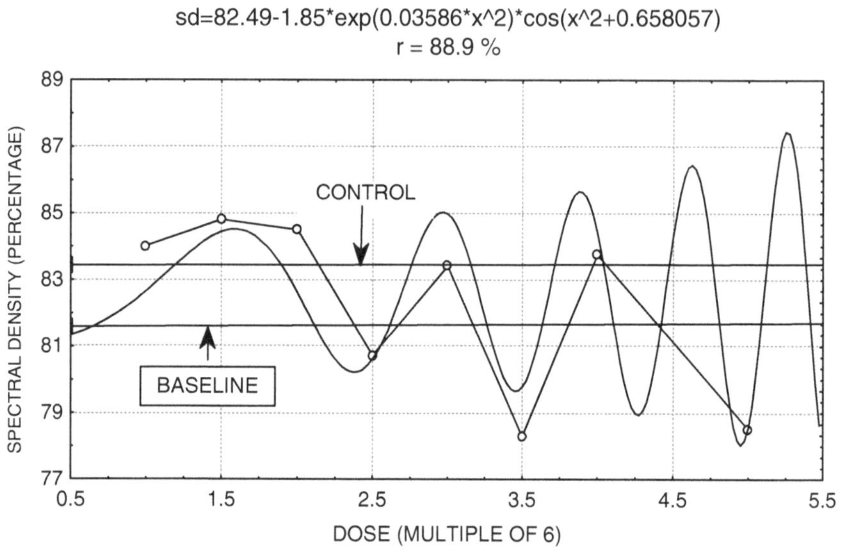

Fig. 5.2a Average values of spectral density in delta band *vs.* dose for ULD-histamine groups. Ultra-Low-Dose as multiple of 6 centesimal shake/dilution steps (First publication at "*Estévez-Delgado and Ruiz-Vega. Modelagem não linear das ultra-diluições. Cultura Homeopática, 16: 31–34, 2006*")

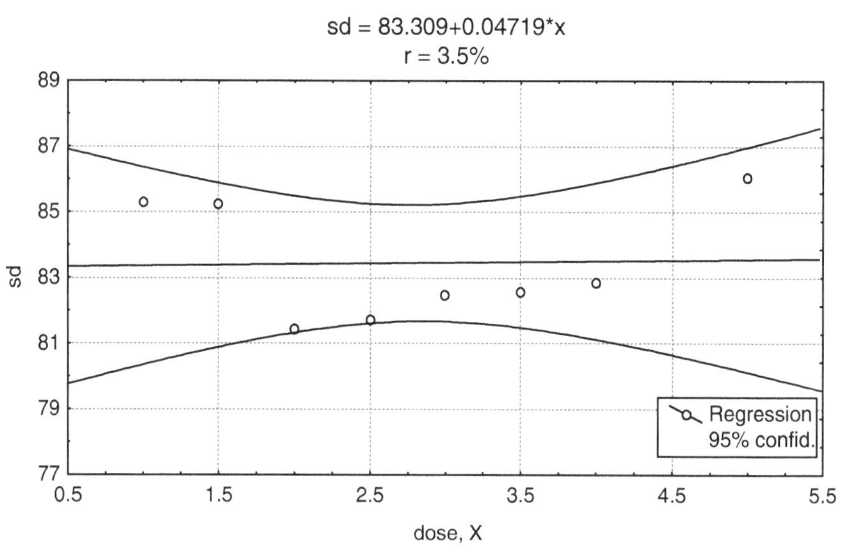

Fig. 5.2b Average values of spectral density in delta band *vs.* dose for controls of ULD groups in Fig. 5.1a. Average values remain practically constant. (First publication at "*Estévez-Delgado and Ruiz-Vega. Modelagem não linear das ultra-diluições. Cultura Homeopática, 16: 31–34, 2006*")

those of group 18; however, values of group 21 are smaller than those of group 24. We only found a statistical difference ($p < 0.05$) from control in group number 21 in 4 hours (4,6,7,8); however post hoc test shows $p = 0.05$ for 15c in the 2 and 7 hour. For group 30c, which corresponds to former results, there is statistical difference in the whole range.

As the dilution increases, ULD histamine groups showed a gradual, but not consecutive, reduction in the average of spectral density in delta band originating an oscillatory pattern after dose equal to 12c (~10^{-24} M); the averages of control remained constant; this behavior becomes more evident plotting average spectral density *vs*. dose (Figs. 5.2a and b).

b. *The relative growth rate, r(x) (Fig. 5.3)*

1. The relative growth rate is a fundamental concept in growth analysis as hazard is in survival analysis. It is defined as the ratio of the growth rate d(y)/dx to achieved growth y (http://www.asu.edu/sas/sugi25/stats/25p277.pdf):

$$r(x) = \frac{1}{y}\frac{dy}{dx}$$

where:

y is spectral density in delta band
x is time

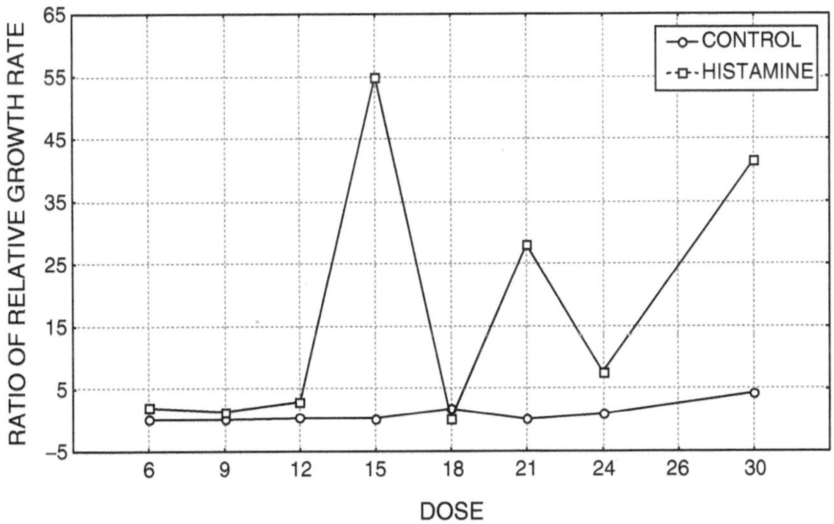

Fig. 5.3 Ratio of relative growth rate, r(x) *vs*. dose. Ratio as quotient of the absolute value of the relative growth rate immediately after dosing over the value during dosing. *Dosing is defined as the number of centesimal shake/dilution steps*. Data for dose 26 was not evaluated. Differences in growth rate appear from dose 15. Considering there are the same conditions, the only source of difference should come from the ULD histamine

Results evidence a flat pattern in spectral density in delta band during the dosing period (the first 2 hours) for histamine groups; immediately after, at t = 3, the spectral density in delta band increases. Control does not display a great difference in this trend. In both cases, this change in tendency is evaluated by means of the ratio *r(2 – 3)/r(1 –2)*, where:

r(1 –2) is the absolute value of growth rate when dosing
r(2 –3) is the absolute value growth rate immediately after dosing

Discussion

Dose-response models for receptor-mediated events should use information on the quantitative relationships between ligand concentration, receptor occupancy, and biological response. Even though histamine is a nice example of ligand-receptor activity (Goodman and Gilman, 1996), according to conventional theory it is difficult to explain the results following traditional Clarck4s approach due to two main issues of the test at hand, i.e. histamine does not cross blood brain barrier (Deli et al., 2000), and the extremely low concentration of the oral dose. Lacking in molecular-based dose metrics, we associate with each dilution a perturbation of unknown characteristics, perhaps of electromagnetic nature. Considering the two former approaches:

a. *The absolute value of the marker*

We did not find correlation between mean values for control and dose; in contrast, experimental data for ULD histamine shows a strong correlation with the equation of an under damped oscillator.

If we focus only on the first 2 hours, assuming that the reduction in the marker is caused by the stimulus during the dosing period, the correlation with the oscillator equation increases to almost 96%. Significant differences remain only in 21 and 30c.

Although this equation comes from a mechanical example, i.e.: pendulum oscillations, the same mathematical principle applies to a whole host of physical systems, among others, acoustic, atomic and electrical, the latter being the most important of non-mechanical applications. Equivalence between mechanical and electrical quantities identifies the displacement as the charge.

b. *Relative growth rate*

Results evidence a flat pattern in spectral density in delta band during the dosing period (the first 2 hours) for histamine groups; immediately after, at t = 3, the spectral density in delta band increases. Control does not display a big difference in this trend.

Ratios for control do not show a great difference in growth rate during and after dosing, i.e. there was not a superimposed perturbation attributable to the solvent, apart from the mechanical sleep disturbance caused by dose administration. In other words, there was no change in their tendencies. Ratios for ultra-diluted histamine also show no difference in tendency up to dose = 12, which approximately is equal to 10^{-24} M (ten to the power of minus 24 molarities), meaning no system response to dosing; however, an oscillation appears beyond this threshold.

Response Selection

Is This the Right Marker?

From our perspective, tracking in time the behavior of the system under stimulus is the most accurate way to find the right response, because it is not only a matter of arriving at some stationary state, but it is also important to assess how this state is reached. The reasoning behind our decision was to track the system's response without adding any extra disturbances. Electromagnetic signals can be good candidates to do this, even more so because they were recorded during sleep, subject awareness is minimized, and it is no possible to "induce" a selected response.

Checking the trend for all groups in Fig. 5.1a, we realize that the homeostatic response drives the system to increase the spectral density in delta band. However, it is possible to recognize homeostasis during the process? According to the protocol, the system evolves between two different homeostatic states; both of them attained spontaneously and recognized conventionally (Recordati and Bellini, 2004):

a. The conscious state of quiet wakefulness through the first 2 hours, when the subject was disturbed by dosing every 20 minutes; hence, the sleep cycle was prevented from evolving, in spite of the animal's being at the beginning of its circadian sleep cycle.
b. The unconscious stable state of NREM sleep in the last 3 hours, which is asymptotically reached. Because of this feature, it is identified as a globally stable and attractive state.

According to the protocol, the system evolves between the two states described above: from the state of resting wakefulness (because of sleep disruption to give the stimulus), until it reaches the NREM stage. In other words, the EEG traces becomes wider than at the beginning of the process, and the prevalence of delta waves increases, thereby reflecting the presence of NREM. Prevalence of delta waves is characterized not only by the spectral analysis but also by the amplitude of the EEG pattern. In fact, at the end of the record, the EEG traces have become wider than at the beginning, reflecting the predominance of NREM.

Self-Organized Critically

Complexity is inherent and intrinsic to living beings; however, it is possible to find a common feature that catches all the diversity and complexity; this common feature is self-organization. The expression Self-Organized Critically (SOC) has been introduced, to describe the tendency of dissipative systems to drive themselves to a critical state, which is identified as an attractor for the dynamics. Equation $y = C * x^{\alpha}$, corresponds to the general form of the mathematical relations known as "power laws"; their main feature is self-similarity or scale invariance (Schroeder, 1991), which in turn is associated with fractality. The fractal concept was coined to describe irregular geometric forms that lack a characteristic (single) scale of length, but can also be applied to self-similar phenomena evolving in time (Bak et al., 1987; Bak and Creutz, 1994).

Fractal physiology as "*adaptive processes which guarantee to respond to unpredictable stimuli and stresses*" is widely accepted (Goldberger, 2001). When certain final state is specific for a given system in determined circumstances, no matter the initial conditions, the tendency to reach such a steady state is termed "equifinality" (Recordati and Bellini, 2004). Because from different initial conditions all the experimental groups evolved to reach the same equifinal state, (circled in Fig. 5.1b), it is assumed that the system has been self-organized in time to reach a global attractor, which in turn implies fractality. The adaptability of response is crucial to confer robustness under unexpected circumstances. Robustness and fractality are generic characteristics of SOC (Carlson and Doyle, 2001), it is not unreasonable to think of homeostasis as a fractal process for restoring the dynamics of health.

Metrics

Response

This is a controversial issue due to the tiny differences among the groups (we only have significance in differences from control in 30c, 21c (4,6,7,8). *Post hoc* test for the two first hours in 15c have a significance equal to 0.05, this is we almost reach statistical significance, which could be solved by increasing the number of subjects, however, the test becomes much more time-consuming and the feasibility of the protocols diminishes. Appropriate facility is also important; to guarantee that the chosen response comes only from the stimulus, it is necessary to isolate the system from any kind of electromagnetic disturbance (net, fm, am, etc.). All the recording cages have Faraday insulation.

Dose

Metrics in dosing is still a problem, so we only refer to it as a "stimulus" that shares some kind of "information" with the system. At this point, the electrical analogy of the model is useful to define the nature of the "information". From the empirical model, the equivalence between mechanical and electrical quantities identifies the displacement (spectral density) as the charge. Because the spectral density was evaluated over 1 hour, this is equivalent to the flow charge, meaning the electrical current, in the time unit (1 hour).

At the same time, the neurophysiologic and biophysical basis of the EEG generation, states that the EEG measures the potential within the voltage fields, generated by dipoles within the cortex, which in turn are due to electrical current flowing into and out the cortical neurons. Such electric activity is primarily due to synaptic activity rather than action potentials (Niedermayer and Lopes da Silva, 1993). Then, from the electric analogy, the marker for the response can be associated with a measure of electric activity, meaning electric current, which in turn is also identified in the EEG generation. This comparison suggests in dose-response some kind of electromagnetic interaction that leads to consider the synaptic activity as the dose target and the response associated with electromagnetic activity.

Stochastic Resonance and ULD Effect

Experimental and epidemiological data have brought forth a discussion about the biophysics of interactions of extremely low frequency electromagnetic fields, ELF EMF. According to this, the numerous structures of living systems can add inputs over time and space; hence a lot of receptors of a particular type on a single cell may send the same message inside the cell, where they may be added. Similarly, repetition of the same message sent over a certain time may be cumulative. In contrast, the noise or random fluctuations may differ at different receptors and partially cancel the signal; still, the averaging of noise over time or number of receptors can substantially improve the capacity of the system to detect small signals (http://www.niehs.nih.gov/emfrapid/html/WGReport/Chapter48.html). Engineers now add noise to some systems, in order to improve how humans perceive the signals. These systems include audio compact discs, analog-to-digital devices, video images, schemes for visual perception and cochlear devices. Noise can also improve human tactile response, muscle contraction and coordination.

The enhancement of the self-response in a system is explained by means of Stochastic Resonance, SR, phenomenon, which can be explained as a cooperative nonlinear phenomenon wherein the signal-to-noise ratio (SNR) at the output of a noisy nonlinear system, suitable to be in two stable states, driven by a weak deterministic modulating signal, assuming to be time periodic, can actually be enhanced by increasing the noise. A complete mathematical model could be found in reference (Millonas, 1996).

On this regard, we have already proposed SR, as a mechanism to explain the biological effect of high dilutions systems by means of the enhancement of system response (Ruiz-Vega et al., 2003) (Table 5.1):

Table 5.1 Similarities between SR phenomenon and ULD effects. ULD stimulus would propitiate noise enhancement through a "constructive interference" (Hahnemann, 1996), which means to add point by point, the signal that would be brought by ULD stimulus to internal noise. In order to get an enhancement, two signals ought to match perfectly

SR phenomenon	ULD effect
System	Living being
States	Healthy or sick
Weak deterministic modulation	Internal autonomous biological cycles
Noise	Noise in nervous system

- The non linear system, complex system, would be the living being, which could be in "healthy" or "sick" states.
- Living beings exhibit diverse kinds of internal cycles at different levels and frequencies, i.e. membrane potential, metabolic oscillations, enzymatic reactions, protein synthesis, epigenetics oscillations, cellular communication by pulses, hormonal signaling, among others (Montero and Moran, 1992). Hence, the weak deterministic modulation is the self-response of the system, which could be enhanced via noise.
- Noise is ubiquitous in nervous systems of living beings. We suggest that certain kinds of noise (in moderate amounts) are related to sick or healthy states, where the so called "pink noise" (Schroeder, 1991) would be associated to healthy state, as a resemblance of *"... all the parts of the living being in a harmony"*(Hahnemann, 1996); total disorder could be related to white noise. Internal noise features would be modified in a sickness condition: the symptoms associated to each malady would imprint to it specific characteristics that could be related with the term *'terrain"* of the French treatises (Vannier, 1991).

The complete hypothesis would be as follows:

Living beings could be oscillating in two meta-stable states: health or sickness. When a healthy subject is perturbed and moved from the healthy to sick state, symptoms would appear at different levels modifying internal noise characteristics. Simultaneously, cyclic signals, devoted to return the system to healthy state, would be self-emitted by the autonomous system. ULD promotes the enhancement, via noise, of this auto-initiated system response promoting its self-recovery. Adaptive Systems could learn to add an optimal amount of noise.

Conclusion

Results from this experiment suggest that the ultra-diluted histamine induces a transient sleep deprivation. The dose-response curve follows, through the range dose tested, an oscillatory pattern that closely match with an under damped oscillator model.

In spite of the fact that empirical models generally do not have a straight link with the underlying mechanism, they can be analyzed in light of available information and provide qualitative insights into the mechanism of the modeled response. Although further replication is needed to set a sound basis on the matter, the usefulness of the empirical model resides in its application as a tool to fit experimental data and make cross-comparisons with other repeating tests. According to this, repetitiveness is essential for modeling, not only the selection of the right response but also the degree of precision to measure the dose and response. There is a great challenge in the field of ULD systems.

Acknowledgements This research was supported by a grant from the scientist research council from the University of Michoacán.

Guadalupe Ruiz-Vega thanks Leticia Pérez O. and Griselda Pérez O. for their help in technical procedures. Thanks for *Cultura Homeopáica* for authorizing the integral reproduction of Figs. 5.2a and b.

References

Bak P, Creutz M. Fractals and self organized criticality. In: Bunde A and Havlin S (Eds.). Fractals in Science. Berlin: Springer-Verlag; 1994.

Bak P, Tang C, Wiesenfeld K. Self-organized criticality: An explanation of 1/f noise. Phys. Rev. Lett.; 1987, 59:381–384.

Carlson JM and Doyle J. Complexity and robustness. Colloquium of The National Academy of Sciences. March 23–24 2001; New York.

Deli MA, Németh L, Falus A, Ábrahám CS. Effects of N,N-diethyl-2-[4-(phenylmethyl) phenoxy]ethamine on the blood-brain barrier permeability in the rat. Eur. J. Pharm.; 2000, 387:63–72.

Goldberger AL. Fractal dynamics in physiology: Alterations with disease and aging. Colloquium of The National Academy of Sciences. March 23–24 2001; New York.

Goodman L and Gilman A. The Pharmacological Basis of Therapeutics. New York: Mc Graw Hill; 1996.

Hahnemann S. Organon of Medicine. Baline, WA: Cooper Publishing; 1996.

Millonas M (Eds.). Fluctuation and Order. New York: Springer-Verlag; 1996.

Montero F, Moran F. Biofísica: Procesos de autoorganización en Biología. Madrid: Eudema Universidad; 1992.

Monti JE. Involvement of histamine in the control of the waking state. Life Sci.; 1993, 53:1331–1338.

Niedermayer E and Lopes Da Silva F. Electroencephalography, Basic Principles, Clinical Applications and Related Fields. Baltimore, MD: William & Wilkins; 1993.

Paxinos G and Watson CH. The Rat Brain in Stereotaxic Coordinates, 2nd edition. New York: Academic Press; 1986.

Poitevin B. Introducción a la Homeopatía: Bases experimentales y científicas. Mexico City: Nueva Editorial Homeopática Mexicana S.A.; 1992.

Recordati G and Bellini TG. A definition of internal constancy and homeostasis in the context of non-equilibrium thermodynamics. Exp. Physiol.; 2004, 89:27–38.

Rechtschaffen A, Kales A (Eds.). A Manual of Standardized Terminology, Techniques and Score System for Sleep Stages of Human Subjects. Bethesda, MD: US Department of Health. Education and Welfare; 1968.

Ruiz G, Torres JL. Homeopathic effect on the sleep pattern of rats. Br. Hom. J.; 1997, 86:201–206.

Ruiz G, Torres JL, Michel O, Navarro R. Homeopathic effect on heart rate variability. Br. Hom. J.; 1999, 88:106–111.

Ruiz-Vega G, Pérez-Ordaz L, Proa-Flores P, Aguilar-Díaz Y. An evaluation of *Coffea cruda* on rats. Br. Hom. J.; 2000, 89:122–126.

Ruiz-Vega G, Pérez-Ordaz L, León-Huéramo O, Cruz-Vázquez E, Sánchez-Díaz N. Comparative effect of *Coffea cruda* potencies on rats. Homeopathy; 2002, 91:80–84.

Ruiz-Vega G, Perez Ordaz L, Cortez Galvan L, Juarez-G.FM. A kinetic approach to cafeinne-*Coffea cruda* interaction. Homeopathy; 2003, 92:19–29.

Ruiz-Vega G, Poitevin B, Perez-Ordaz L. Histamine at high dilution reduces spectral density in delta band in sleeping rats. Homeopathy; 2005, 94:86–91.
Sainte-Laudy J, Belon P. Inhibition of human basophil activation by high dilutions of histamine. Agents Actions; 1993, 38:C245–C247.
Schroeder M. Fractal, Chaos, Power Laws. New York: W.H. Freeman; 1991.
Schwierin B, Borbély AA, Tobler I. Effects of N6 -cyclopentyladenosine and caffeine on sleep regulation in the rat. Eur. J. Pharm..; 1996, 300:163–171.
Tuomisto L. Involvement of histamine in circadian and other rhythms. In: Watanabe T, Wada H (Eds.). Histaminergic neurons: Morphology and function. Boca Raton, FL: CRC Press.; 1991.
Vannier L. La Práctica de la Homeopatía, Chap3, pp.169–360, México, Editorial Porrúa, 1991.

Chapter 6
Hepatic Cell Growth Models for the Study of Ultra High Dilutions

Kátia Silva Martinho[2], Vanessa Del Bianco de Bento[2], Graziella Ulbricht Benvenga[2], Viviane Aparecida Marcondes[2], and Leoni Villano Bonamin [1,2]

Introduction

The regenerative hepatic process has been of interest since Ancient Greek time, according to the Prometheus myth. Nowadays, hepatic regeneration study is incontestable due to its clinical importance. For all the characteristics mentioned below, studies involving hepatic regeneration and hepatic carcinogenesis are a good basic research model to study biological effects of ultra high dilutions in living organisms.

It is well known the majority of the liver is made from hepatocytes, classified as "stable cells", which means that under normal conditions, they have low "turn over" or low renew rate, with mitotic activity up to 1% (Cotran et al., 2004). However, when a reduction of functional liver parenchyma occurs, due to a lesion or partial ablation of this organ, the proliferation activity increases until 100% of its parenchyma is regenerated. Many serious hepatic diseases such as cirrhosis and hepatocellular carcinoma are a disturbance result of the regenerative process. In the last years, much effort has been dedicated to understand the molecular mechanisms coordinating these processes.

The partial liver resection is a classical research model for the study of hepatic regeneration. After 24–48 hours after partial hepatectomy, approximately 10% of the actual parenchyma is into the "S" stage of the cellular cycle, which means DNA synthesis process (Cotran et al., 2004). Therefore the beginning of the cellular regeneration is linked to nuclear transcription protein sequence expression and cellular cycle regulators proteins such as Fos, Jun, Myc, p53, NFkB and cyclines (Cotran et al., 2004).

When the hepatic parenchyma is completely regenerated, the cells in proliferation return to being quiescent. Many extra-cellular factors control this sequence of events, mostly growth factors and cytokines. The HGF (Hepatocyte Growth Factor)

[1] Universidade Paulista. São Paulo, Brazil. leonibonamin@gmail.com

[2] Universidade de Santo Amaro. São Paulo, Brazil

Presented at the XVIth Symposium of GIRI, Monaco, 2002

L.V. Bonamin (ed.) *Signals and Images*,

is made by hepatic stromal cells and induce proliferation in the hepatocytes and adjacent cells (Fausto, 2000). O EGF (Epidermal Growth Factor) is a mitogenic factor with action on the first stages of hepatic regeneration while the TGF-α (Transforming Growth Factor) has an important modulation action on late stages of regeneration process. The main cytokines involved in the process are respectively Il-6 and TNF-α, both produced by resistant macrophages. They all induce the reminiscent hepatocyte priming, in other words, they make the cells responsive to growth factors. To complete the growth process, the TGF-β (Transforming Growth factor β) produced by hepatic stroma is released when the regeneration is complete (Cotran et al., 2004).

Concomitant with the chemical mediators involved, many cellular components take part in hepatic parenchyma regeneration. The main cells involved are stem cells, localized between hepatocytes and biliary ducts. Stem cells can be differentiated into hepatocytes or ductal epithelium. It is well known that biliary ductal proliferation occurring in a chronic hepatic lesion is a consequence of stem cell modification to proliferate canalicular components, even in deficit of its own hepatocyte regeneration (Thorgeirsson, 1996). Additionally, the star cells, also known as Ito cells, play an important role producing matrix protein (collagen and laminin) in a normal liver, and also re-organizing the hepatic microcirculation (Dagli, 1994; Dagli et al., 1998).

Furthermore, hepatocarcinogenesis is largely used as an experimental model to study oncogenesis and neoplastic development. Characteristics like the organ size, the cancer induction susceptibility, the well known biochemistry aspects and cellular proliferation manipulation possibility, make this organ ideal for such studies (Farber and Cameron, 1980).

Classically, the neoplastic development study on urinary bladder, respiratory tract, intestines, mammary gland, central nervous system and others has been possible by the administration of initiators chemical carcinogens (Farber and Cameron, 1980).

Functional and structural changes as result of hepatocarcinogens interaction with cellular structures can chronically establish new cellular patterns with neoplastic development, with or without metastasis (Farber and Sarma, 1987).

Regarding the type of change resulting from carcinogen agents, it is possible classify them in two groups:

- Initiators: cause irreversible changes into the cell genome
- Promoters: cause clone expansion of the initiated cells after carcinogen exposition

The Resistent Hepatocyte Model was suggested by Solt and Faber (1976). This model serves three purposes:

- The carcinogen toxicity resistance can be obtained during the carcinogenesis process.
- Many carcinogens are inhibitors of hepatic regeneration.
- Pre-neoplastic hepatocytes show resistance to some cytotoxic characteristics of many carcinogens.

In this model, rats are exposed to a dose of initiator carcinogen called diethylnitrosamine (DEN). After some time, the same rats are exposed to low doses of another initiator called 2-acetylaminofluorene (AAF), and then partial hepatectomy (70% of liver parenchyma) is performed to promote proliferation. In this way, the initiated hepatocytes are stimulated while unaffected cells do not go through regeneration, after hepatectomy. The first pre-neoplastic nodules can be observed around 5–10 days post partial hepatectomy.

Preliminary studies on cellular proliferation and ultra high dilution were developed at the Animal Pathology Laboratory, Santo Amaro University, Brazil. They demonstrate that ultra high diluted dexamethasone modifies cellular migration and exudative process on inflammation and on the interaction between Ehrlich tumor and its host (mice). In addition, dexamethasone is a good experimental model regarding its low cost and its well known mechanism of action (Bonamin et al., 2001).

Dexamethasone exert effects on many physiopathologic processes, such as vascular changes during acute inflammation, proliferation in chronic inflammation and tissue repair, and also on proliferation and immune response during tumor development. This multifactorial characteristic allows the study of ultra high dilutions in various experimental models, facilitating the understanding of ultra high dilutions effect, predominantly for the possibility of generalizing the conclusions obtained from different studies.

As a synthetic steroid, its main intracellular action is modulating genic expression (Bonamin and Paulino, 1999). In 1998, Frod-saleh et al. demonstrated steroids and non steroid drugs inhibiting effects over NF-kappa B. This transcription factor is described as one of the hepatic priming mechanisms during the regeneration process (Fausto, 2000). Therefore the NF-kappa B expression is the conjecture for the study of dexamethasone ultra high dilutions on hepatocarcinogenesis.

A series of trials based on all information above was developed by Bonamin and colleagues to elucidate the understanding of high dilutions effects in living organisms.

Effects of Ultra High Diluted Dexamethasone on Hepatic Proliferation Trials

This study was developed in three stages: hepatic regeneration, hepatocarcinogenesis and histochemical analysis. In the first and second stages, the trial was developed with adult male Wistar rats (N = 10 animals per group), allocated in standard polyethylene cages and controlled circadian cycle (from 6:30 a.m. to 6:30 p.m. light period, and from 6:30 p.m. to 6:30 a.m. dark period), temperature from 19°C to 26°C, feeding *ad libitum*. In the third and last stage, there was no need for using a new sample of animals, since the histochemical analysis was performed on paraffin blocks obtained from the previous stages.

In all stages animals were treated weekly, via subcutaneous injection from the first day after hepatectomy up to 30 days after surgery. The treatment protocol was respectively divided in the following groups.

Group I – Phosphate Buffered Saline (PBS)

Phosphate buffered saline tablet (Sigma) was diluted in purified water and used as a negative control treatment. PBS used as a control treatment and also used on experimental dilutions was made from the same tablet.

Group II – Ultra High Diluted Dexamethasone (Decadron®) 7CH

Dexamethasone 7CH was prepared from six consecutive centesimal (1:100) dilutions of dexamethasone disodium phosphate (Decadron®) in purified water. A seventh and last dilution was made in phosphate buffered saline (PBS) solution, respectively in the same proportion of those previous dilutions. The last solution molarity was equivalent to 10^{-17}M. For each dilution, the glass tube containing the solution was vigorously agitated following classical Hahnemann's method.

Group III – Dexamethasone mix (Ultra High Diluted Dexamethasone 7CH + Dexamethasone)

A mixture of Dexamethasone 7CH prepared as described above with Dexamethasone (Decadron®) at 4 mg/kg dose was prepared. The commercial form of Dexamethasone was diluted into its own high diluted form (Dexamethasone 7CH) in order to reach an optimal concentration capable to set the chosen dose (4 mg/kg).

Group IV – Dexamethasone (Decadron®)

Dexamethasone (Decadron®) was used as a positive control at 4 mg/kg dose. The commercial form of Dexamethasone was diluted into PBS in order to reach an optimal concentration capable to set the chosen dose (4 mg/kg).

Group V – Isopathy (2-acetylaminofluorene 7CH)

This substance is not hydro soluble, consequently the first three dilutions were made by triturating with lactate. The first dilution was made by one part of 2AAF in 99 parts of lactate and again triturated, then, one part of this second dilution was again mixed with 99 parts of lactate, and from then on subsequent dilutions in PBS.

All drugs were prepared on the day of trial or 1 day before and kept under refrigeration.

Bioethical Standard

All trials were conducted according to the "Scientific Animal Vivisection Norms" Brazilian law n°6.638, from 8 May 1979".

Hepatic Regeneration

The first step to promote hepatic regeneration is performing partial hepatectomy. Before the procedure, animals received Ketamine and Xylazine as anaesthetic drugs, in a proportion of 2:1 respectively, and then were submitted to epigastric sagittal-median incision exposing the xiphoid process. Using a forceps and scissor, the xiphoid process was removed and then left lateral liver lobe resection was performed, followed by absorbable suture. Although in the scientific literature this lobe represents 70% of the total liver was observed, it actually represents 30% of the total liver through this trial. After muscle and skin suture, animals were closely observed for 24 hours until total recovery from surgery.

Treatment

In this trial animals were divided in four groups only (Groups I to IV as previously described), they did not received the 5th treatment which is related to carcinogens.

Morphometric Analysis

After 30 days from Hepatectomy, all animals were euthanized and the total liver was removed and weighed. From the largest lobe was removed one central fragment of 30–40 mm width approximately, in a cranio-caudal direction, then fixed in methacarn (60% methanol 30% chloroform 10% glacial acetic acid) for 4 hours, followed by alcohol 70% immersion for 2 days maximum until paraffin inclusion. Slides were stained by HE method for cellular counting.

Five random microscopic fields were chosen and defined under 100x objective lens, with a 200 points reticule. The percentage of points for each structure was counted to define oval cells, megalocytic cells, mitotic figures, sinusoids, necrotic areas and fibrosis area.

Statistical Analysis

Non parametric Qui square (X^2) method was used for cells quantification and ANOVA/Tuckey Krammer methods were used to compare liver weight between groups, being $p \leq 0.05$.

Results

There was no significant result on hepatic regeneration, regarding to liver weight recovery, in animals treated with high diluted dexamethasone. The only effect observed was the decrease in liver mass regeneration followed to dexamethasone 4 mg/kg treatment (Fig. 6.1). No differences were found among groups regarding to cell counting neither (data not shown).

Hepatocarcinogenesis (Bonamin et al., 2002)

The resistant hepatocyte model was used to induce hepatocarcinogenesis. In this model, rats were exposed to repeated doses (3.5 mg/kg) of initiator carcinogen called 2-acetylaminofluorene (AAF) in alternate days, by *gavage* and then on the 20th trial day partial hepatectomy (30% of liver's parenchyma) was performed, as

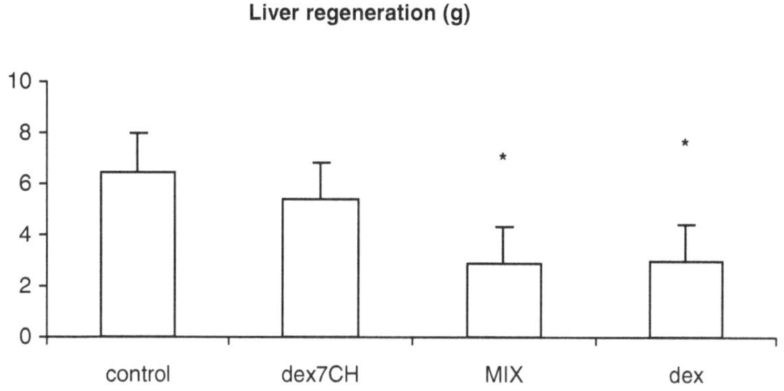

Fig. 6.1 Liver mass regeneration (grams) measured after 30 days from hepatectomy. Four groups of treatments were performed: control – treated with PBS; dex7CH – treated with dexamethasone 7CH; MIX – treated with dexamethasone 4 mg/kg diluted in dexamethasone 7CH; dex – treated with dexamethasone 4 mg/kg diluted in PBS. ANOVA/Tuckey-Krammer, * p = 0.01 in relation to control

proliferation or promoter stimulation. Following hepatectomy, animals continued to receive AAF treatment for 10 days and were euthanized on the 30th trial day (see the framework below).

Alternate days I 2AAF	Day 20 P	Alternate days I 2AAF	Day 30 euthanasia
3.5 mg/kg	Hepatectomy	3.5 mg/kg	Pre-neoplastic
Corn oil	30%	Corn oil	Nodules study
Gavage		Gavage	
→			
Treatment			
Legend			
I: Initiator			
P: Promoter			
DEN: diethylnitrosamine			
2AAF: 2-acetylaminofluorene			

Treatment

In this trial animals were divided in five groups receiving all treatments previously described.

Morphometric Analysis

After 30 days from hepatectomy, all animals were euthanatized, the number of gross pre-neoplastic lesions (PNLs) was counted and the same procedure of biopsy and staining was performed, as described above (see regeneration protocol).

The microscopic phenotype analysis of pre-neoplastic lesions (PNL) was done considering four pre-neoplastic lesion types, respectively:

- *Clear cell P.N.L:* clear and vacuolated cells, few sinusoids, few mitosis
- *Mixed P.N.L:* clear and acidophilic cells, few mitoses, sinusoids more frequent
- *Acidophilic P.N.L:* acidophilic cells, megalocytes, mitoses (typical or atypical) and sinusoids frequent
- *Basophilic P.N.L:* basophilic and atypical cells, mitoses (typical and atypical) frequent

Statistical Analysis

ANOVA/Tuckey-Krammer and Fisher test were used to group comparison in relation to mitosis and macroscopic/microscopic pre-neoplastic nodules counting, being $p \leq 0.05$.

Results

PNLs development got quantitatively and qualitatively (phenotype) worse after dexamethasone 7CH treatment, even in the presence of dexamethasone in pharmacological concentration. Thus, the number of pre-neoplastic nodules and malignant phenotypes were significant higher in both groups treated by Dexamethasone 7CH (Figs. 6.2 and 6.3). Worsening results in relation to malignancy were also shown by a higher number of mitoses, mainly atypical, in the Dexamethasone 7CH treated group (Fig. 6.4).

The high dilution action was preponderant over the pharmacological one, it was shown by more PNLs (all types), significant for acidophilic and basophilic PNL (worsening), in groups treated by dexamethasone 7CH and dexamethasone 7CH mixed with dexamethasone 4 mg/kg. On the other hand, a mild protective effect was observed on the group treated by isopathy (DEN + AAF 7CH) showing less typical mitoses (Fig. 6.4).

Histochemical Study

This trial was designed to further understanding of the results represented above. To investigate meticulously the hepatic structure after treatment, histochemical and morphometrical characterization were observed through the following staining methods: P.A.S (glycogen storage), Prussian blue (demonstrates the blue granules of hemosyderin in hepatocytes and Kupffer cells), modified Glick (biliary pigment retention) and Masson's Trichrome (fibrosis).

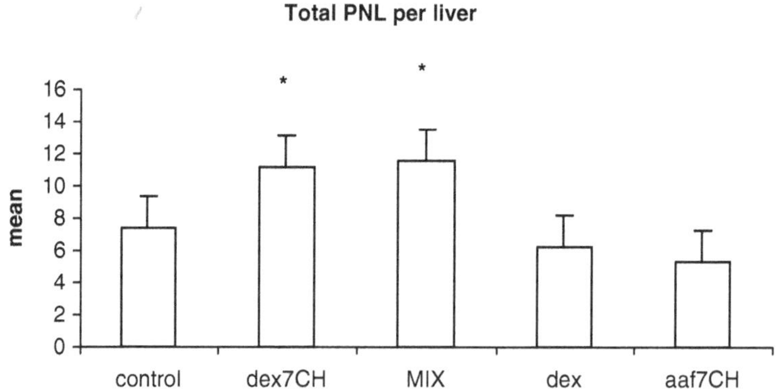

Fig. 6.2 Total number of PNL per liver after 30 days of AAF chronic exposure (10 days from hepatectomy). Five groups of treatments were performed: control – treated with PBS; dex7CH – treated with dexamethasone 7CH; MIX – treated with dexamethasone 4 mg/kg diluted in dexamethasone 7CH; dex – treated with dexamethasone 4 mg/kg diluted in PBS; aaf7CH – treated with AAF 7CH. ANOVA/Tuckey-Krammer, * $p = 0.01$ in relation to control

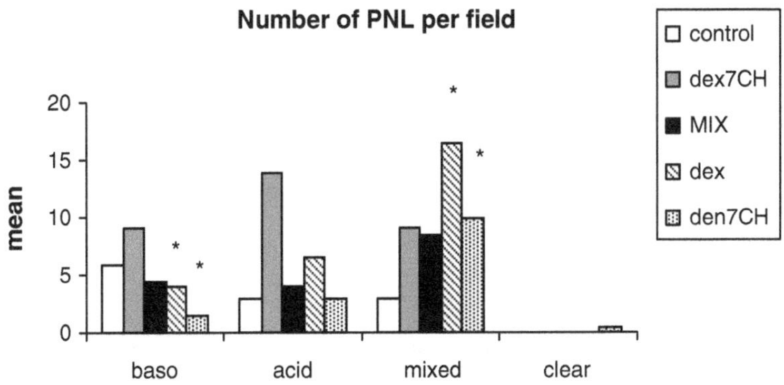

Fig. 6.3 Differential counting of PNL after 30 days of AAF chronic exposure (10 days from hepatectomy). Five groups of treatments were performed: control – treated with PBS; dex7CH – treated with dexamethasone 7CH; MIX – treated with dexamethasone 4 mg/kg diluted in dexamethasone 7CH; dex – treated with dexamethasone 4 mg/kg diluted in PBS; aaf7CH – treated with AAF 7CH. Fisher test, * p = 0.03 in relation to control

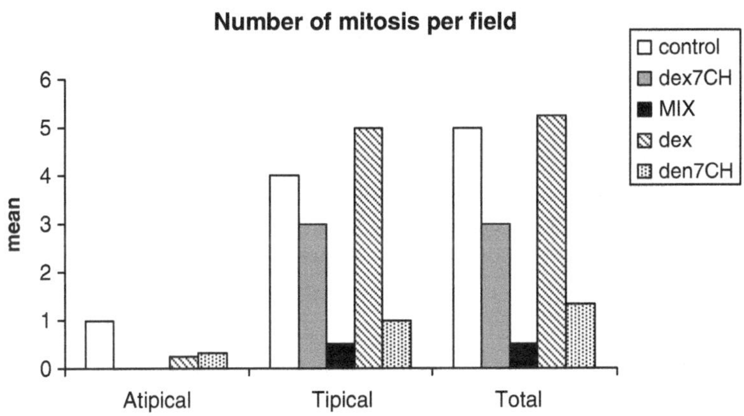

Fig. 6.4 Mitosis counting after 30 days of AAF chronic exposure (10 days from hepatectomy). Five groups of treatments were performed: control – treated with PBS; dex7CH – treated with dexamethasone 7CH; MIX – treated with dexamethasone 4 mg/kg diluted in dexamethasone 7CH; dex – treated with dexamethasone 4 mg/kg diluted in PBS; aaf7CH – treated with AAF 7CH. Fisher test

Method

Histopathological slides from previous studies on resistant hepatocyte model were submitted to specific staining methods. The analysis was performed by two independent observers, on digital microphotography (three fields per slide), designating 1 to 5 scores, in different places and time. For Masson's Trichrome method, quantification was done by Image Tool 3.0 – image analysis software, measuring the total fibrosis area in each animal, summarizing three microphotography fields per slide. All observations were done in blind.

A new sample of animals was not necessary for this trial development, once it was done on archived paraffin blocks from previous studies.

Periodic Acid-Schiff (PAS)

PAS is a common method used in pathology and histology laboratories routine. This method is related to the carbohydrate histochemistry, and easily allocates visualization of molecules high in carbohydrates such as glycogen, mucoprotein and glycoprotein. Glycogen is the main carbohydrate storage source in animals and its identification is often important.

PAS specificity is a reaction derived from the sequential use of two selective reagents, periodic acid (HIO_4) and Schiff's reagent. Periodic acid oxide free hydroxyl groups in two adjacent carbon atoms, like 1.2-glycol in hexoses, or adjacent hydroxyls and amino groups in hexosamine. Hydroxyl groups are converted in aldehydes, the carbon to carbon linking is cleaved, and under P.A.S. conditions, the following oxidation is blocked. Aldehydes then are ready to react to Schiff's reagent creating a stable colored complex. The usual Schiff's reagent is called leucofuchsin, or fuchsin-sulphurous acid, a chromogenic complex which produces a stable red when reacting to aldehydes (Lison, 1960).

Many substances in epithelial and connective tissues react to P.A.S. method, mostly: glycogen, epithelial mucin, golgi apparatus, glycocalyx, basal membranes, and polysaccharide protein complex on amorphous fundamental substance in various connective tissues.

Method description:

1. Remove paraffin
2. Wash in purified water for 5 minutes
3. Periodic acid for 5 minutes
4. Wash in purified water (three times)
5. Filtered Schiff's reagent for 15 minutes
6. Wash in water for 10 minutes
7. Stain by Harris' hematoxylin for 2 minutes
8. Wash in purified water
9. Dehydrate and make up the slide

Glick's Reaction

The reaction to show evidence of bilirubin is based on the biliverdin oxidation, by appropriate reagents: iodine, sodium hyposulphite and acetic saphranin (Tolosa et al., 2003).

Method description:

1. Remove paraffin
2. Iodine for 12 hours (overnight)
3. Wash in purified water for 10 minutes
4. Discolor by hyposulphite for 10 minutes
5. Stain by acetic saphranin for 3 minutes
6. Wash in alcohol 70%
7. Dehydrate and make up the slide

Prussian Blue Staining

Created by Perls in 1966, iron can be verified on ferric ferricyanide (or Prussian blue): $[Fe\ (CN)_6]_3Fe_4$. It is an extremely sensitive reaction to ferric salts, but not to ferrous salts.
Method description:

1. Remove paraffin
2. Wash in purified water
3. Drop on the slide one part of potassium ferrocyanate and three parts of chloridic acid 1% made at the time
4. Warm up the solution in a test tube at 70°C for up to 60 seconds (observe vapor)
5. Wash in purified water
6. Eosin stain for 5 minutes
7. Wash in purified water
8. Dehydrate, and make up the slide

Masson's Trichrome

This method is based on four reagents: acid fuchsin, hematoidin, picric acid, and phosphomolybdic acid. The objective is visualizing collagen, stained in blue and other fibrillar structures.
Method description:

1. Remove paraffin
2. Wash in water for 1 hour
3. Wash in purified water for 5 minutes
4. Stain by Harris' hematoxylin for 5 minutes
5. Wash in water for 5 minutes
6. Wash in purified water for 5 minutes
7. Stain by acid fuchsin for 5 minutes
8. Wash in purified water for 10 minutes
9. Phosphotungstic acid for 10 minutes

10. Stain by aniline blue for 10 minutes
11. Wash in purified water for 5 minutes
12. Wash in acetic acid 1% for 3 minutes
13. Wash in purified water for 3 minutes
14. Dehydrate, and make up the slide

Statistical Analysis

The statistical method used was Kruskal-Wallis/Dunn, being $p \leq 0.05$.

Results

Animals treated with dexamethasone 7CH combined with dexamethasone 4 mg/kg presented significant increase in the biliary stasis in relation to those treated with dexamethasone 4 mg/kg. Hepatic glycogen was higher in the combined dexamethasone group, statistically significant in comparison to all other groups. Prussian Blue staining has shown weak positivity related to non soluble reduced iron in hepatic tissue. There was no difference related to iron pigmentation and fibrosis observed under Prussian Blue and Masson's trichrome staining (Table 6.1).

Discussion

PNLs can be affected by modulation in its phenotypic characteristics, as observed *in vivo*, but also show important architectural changes observed by histochemical analysis. It is well known that dexamethasone blocks hepatic proliferation after hepatectomy, but induces temporary hepatic hypertrophy (Nagy et al., 2001). Bartolozzi et al. (2007) identified by image diagnosis hypervascularization and cirrhosis-like histological changes in early development stages of hepatocarcinomas.

Table 6.1 Scores of specific staining of liver histopathological samples after 30 days of AAF chronic exposure (10 days from hepatectomy). Five groups of treatments were performed: control – treated with PBS; dex7CH – treated with dexamethasone 7CH; MIX – treated with dexamethasone 4 mg/kg diluted in dexamethasone 7CH; dex – treated with dexamethasone 4 mg/kg diluted in PBS; aaf7CH – treated with AAF 7CH. Kruskal-Wallis/Dunn, $^{*}p \leq 0.01$ in relation to the other groups. Data represented by median and interval

Groups/ treatments	Control	Dex7CH	MIX	Dex	aaf7CH
PAS	2 (1–2.5)	1.5 (1–2)	3* (1.5–4)	1 (1–2)	1 (1–2)
Glick's reaction	3 (1–4)	2 (1–3)	3* (1–5)	2 (1–4)	3 (1–4)
Prussian blue	1 (1–2)	1 (1–2)	1 (1–2)	1 (1–2)	1 (1–2)
Masson's Trichrome	1 (1–5)	1 (1–3)	1 (1–4)	1 (1–3)	1 (1–3)

The main disorders among regular histophysiological changes in PNL development are alterations to oval cell proliferation and communication between GAP junctions (Dagli et al., 2004; Roskams et al., 2003; Kuhlmann and Peschke, 2006).

Perhaps hepatic hypertrophy and/or oval cells proliferation might represent anatomic substrates which justify observed cholestasis in animals with higher number of nodules. Nevertheless, different hypothesis must to be considered, such as possible functional alterations. This kind of alteration is described by Monte et al. (2005), who observed biliary acids transporters expressions changes on hepatocarcinogenesis development.

Hepatic glycogen was higher in mixed dexamethasone group, statistically significant in comparison to all other groups. It shows again a strong connection between total PNLs incidence and histometric scores. In fact, it is known that a higher glycogen amount can be observed in acidophilic nodules (Sorof and Custer, 1987).

Once more, the results point to a possible potent effect between both dexamethasone types combined, since corticosteroids have a physiologic function of stimulate hepatic glycogen synthesis, which can be increased during proliferation process (Assy and Minuk, 1997).

The lack of variation on peri-portal fibrosis development reinforce the hypothesis that higher cholestasis observed in combined dexamethasone group, actuality has a link with hepatocyte metabolic alteration or oval cell proliferation. Such metabolic alterations probably do not interfere in the iron metabolism and ferritin availability, since hemosyderosis variations nor venous congestion and red cells degradation by Kupffer cells were not observed in this study.

The non soluble reduced iron in hepatic tissue stained by Prussian Blue demonstrates saturation and ferritin structural modification with consequent hemosyderin production. This phenomenon is very common when hemolysis and blood stasis are present, with hemoglobin decomposition or in cases of higher iron intestinal absorption (Brito, 1999). In this case, the wake positivity founded shows that the intestinal iron absorption do not appeared an important factor on carcinogenesis process in this experimental model, do not matter which treatment was used.

Conclusion

Although no ultra high diluted dexamethasone effect can be seen in normal liver regeneration, it can lead to undesirable effects on pre-neoplastic lesion development, in contrast to a partial protection given to the host by isopathic treatment considering some phenotypic features, as mitosis index.

The present data are convergent; suggesting that ultra high diluted dexamethasone (10^{-17}M) prepared by Hahnemann's method modulates pre-neoplastic evolution in rats. Thus, for further application it is important to observe that high dilution of endogenous-like substances must be used carefully in animals bearing malignant or pre-malignant processes.

References

Assy N, Minuk GY. Liver regeneration: methods for monitoring and their applications. J. Hepatol.; 1997, 26: 945–952.

Bartolozzi C, Crocetti L, Lencioni R, Cioni D, Della Pina C, Campani D. Biliary and reticuloendothelial impairment in hepatocarcinogenesis: the diagnostic role of tissue-specific MR contrast media. Eur. Radiol.; 2007. Eletronic version (SpringerLink).

Bonamin LV, Paulino CA. Imunofarmacologia. In: Spinoza HS, Gorniak S, Bernardi MM (Eds). Farmacologia Aplicada à Veterinária. 2o Ed. Rio de Janeiro: Guanabara Koogan Editora; 1999.

Bonamin LV, Martinho KS, Nina AL, Caviglia F, Do Rio RGW. Very high dilutions of dexamethasone inhibit their own pharmacological effects *in vivo*. Br. Hom. J.; 2001, 90: 198–203.

Bonamin LV, Martinho KS, Marcondes VA, Benvenga GU. Ultra-High Dilutions and proliferative activity of liver. Annals of the GIRI meeting. October, 2002; Monte Carlo, Monaco.

Brito T. Pigmentos e pigmentação patológica. In: Montenegro MR, Franco M. Patologia – Processos Gerais. São Paulo: Atheneu; 1999.

Cotran RS, Kumar V, Collins T. Pathologic Basis of Disease – Robbins. 7th Ed. Philadelphia, PA: W.B. Saunders; 2004.

Dagli MLZ. Aspectos Morfológicos dos fígados de ratos tratados com beta-caroteno ou vitamina A e submetidos ao modelo do "Hepatócito Resistente" de Carcinogênese. Tese de doutorado. São Paulo: FMVZ-USP, Brasil; 1994.

Dagli MLZ, Guerra JL, Sinhorini IL, Wu TS, Rizzi MBSL, Penteado MVC, Moreno FS. Beta-carotene reduces the ductular (oval) cell reaction in the liver of Wistar rats submitted to the resistant hepatocyte model of carcinogenesis. Pathology; 1998, 30: 259–266.

Dagli ML, Yamasaki H, Krutovskikh V, Omori Y. Delayed liver regeneration and increased susceptibility to chemical hepatocarcinogenesis in transgenic mice expressing a dominant-negative mutant of connexin 32 only in the liver. Carcinogenesis; 2004, 25: 483–492.

Farber E, Cameron R. The sequential analysis of cancer development. Adv. Cancer Res.; 1980, 31: 125–226.

Farber E, Sarma DSR. Hepatocarcinogenesis: A dynamic cellular perspective. Laboratory Investigation; 1987, 1: 4–22.

Fausto N. Liver regeneration. J. Hepatol.; 2000, 32: s.1, 19–31.

Kuhlmann WD, Peschke P. Hepatic progenitor cells, stem cells, and AFP expression in models of liver injury. Int. J. Exp. Pathol.; 2006, 87: 343–359.

Lison L. Histochimie et Cytochimie Animales. Principes et Méthodes, vol. I, vol. II. Paris: Gauthier-Villars Ed.; 1960.

Monte MJ, Fernandez-Tagarro M, Macias RI, Jimenez F, Gonzalez-San Martin F, Marin JJ. Changes in the expression of genes related to bile acid synthesis and transport by the rat liver during hepatocarcinogenesis. Clin. Sci.; 2005, 109: 199–207.

Nagy P, Teramoto T, Factor VM, Sanchez A, Schnur J, Paku J, Thorgeirsson SS. Reconstitution of liver mass via cellular hypertrophy in the rat. Hepatology; 2001, 33: 339–345.

Roskams TA, Libbrecht L, Desmet VJ. Progenitor cells in diseased human liver. Semin. Liver Dis.; 2003, 23: 385–396.

Solt D, Faber E. New principle for the analysis of chemical carcinogenesis. Nature; 1976, 263: 701–703.

Sorof S, Custer RP. Elevated expression and cell cycle deregulation of a mitosis-associated target polypeptide of a carcinogen in hyperplastic and malignant rat hepatocytes. Cancer Res.; 1987, 47: 210–220.

Thorgeirsson SS. Hepatic stem cells in liver regeneration. FASEB J.; 1996, 10: 1249.

Tolosa EMC, Rodrigues CJ, Behmer AO, Freitas Neto AG. Manual de Técnicas para Histologia – Normal e Patológica. 2ª Edição. Barueri: Manole; 2003.

Chapter 7
Models with Plants, Microorganisms and Viruses for Basic Research in Homeopathy

Lucietta Betti[1], Grazia Trebbi[1], Daniele Nani[2], Vera Majewsky[3], Claudia Scherr[4], Tim Jäger[3,5], and Stephan Baumgartner[4,5]

Introduction

Most criticism about homeopathy concerns the lack of a scientific basis and theoretical models. In order to be accepted as a valid part of medical practice, a well-structured research strategy for homeopathy is needed. This is often hampered by methodological problems as well as by gross underinvestment in the required academic resources. Fundamental research could make important contributions to our understanding of the homeopathic and high dilutions mechanisms of action.

Plant- and microorganism-based experimentation appears suitable to this goal, making it possible to overcome some of the disadvantages of clinical trials: botanical and microbial trials do not present neither placebo effect nor ethical problems, and rely on a very cheap and almost inexhaustible source of biological material (Betti et al., 2003a). Moreover, relatively simple model systems can be adopted so that a more direct treatment/effect relationship and large data samples for structured statistical analyses can be obtained. This is a very important feature because it allows a large number of experimental repetitions and external replications to be performed, useful for studying the problem of irreproducibility so often reported in homeopathic literature (Steffen, 1984; Baumgartner et al., 1998; Binder et al., 2005). In fact, one of the major challenges of homeopathic fundamental research today is the reproduction of preclinical studies that have shown significant effects of ultra highly diluted substances compared to control groups. The lack of reproducibility represents a crucial difficulty in testing homeopathy and has stimulated explanations of homeopathic

[1] Bologna University, Bologna, Italy, lucietta.betti@unibo.it

[2] Italian Society of Anthroposophical Medicine, Milan, Italy

[3] Research Institute of Organic Agriculture (FiBL), Frick, Switzerland

[4] Institute Hiscia, Arlesheim, Switzerland

[5] University of Bern, Bern, Switzerland

Presented part at the XVIth Symposium of GIRI, Monaco, 2002 and part at the XVIIth Symposium of GIRI, Graz, 2003

treatment effects using complexity theory (Hyland and Lewith, 2002). On the basis of the experimental evidences in wheat and tobacco models (Betti et al., 2003b; Brizzi et al., 2005), a recent proposed hypothesis is that the systematic reduction of variability might be one of the peculiar actions of ultra high dilutions (Nani et al., 2007). Furthermore, since the main cell structures and functions are common in the majority of eukaryotes (Lam et al., 2001; Carrington and Ambros, 2003), plant and eukaryotic microbial bio-assays could be of interest also from a medical point of view, at least as complementary to clinical studies.

Since the pioneering works of Kolisko on wheat germination (Kolisko, 1923) and Junker on growth of microorganisms (paramecium, yeast, fungi) (Junker, 1928), a number of experiments have been performed either with healthy organisms (various physiological aspects of growth) or with artificially diseased organisms, which may react more markedly to homeopathic treatments than healthy ones. In the latter case, the preliminary stress may be either abiotic, e.g. heavy metals, or biotic, e.g. fungal and viral pathogens or nematode infection. Research has also been carried out into the applicability of homeopathic principles to crop growth and disease control (agro-homeopathy): because of the extreme dilutions used, the environmental impact is low and such treatments are well suited to the holistic approach of sustainable agriculture (Betti et al., 2006). Unfortunately, as Scofield reported in an extensive critical review (Scofield, 1984), there is little firm evidence to support the reliability of the reported results, due to poor experimental methodology and inadequate statistical analysis. Moreover, since there is no agricultural homeopathic pharmacopoeia, much work is required to find suitable remedies, potencies and dose levels.

Recently, high methodological standards have been applied to basic research into homeopathy with different plant and microbial model systems: external influences (such as light, temperature, humidity, soil and seed quality) have been considered; most handling steps and instrumental measurements have been carried out blind in order to exclude unconscious influences by the researcher; experimental designs have been guided by adequate statistical standards.

Herein, only subsequent literature published after Scofield's critical review (Scofield, 1984) will be considered. Therefore, the present overview is divided in 4 sections:

1. Models based on healthy plants, microorganisms, and viruses
2. Models with impaired plants and microorganisms (abiotic stress)
3. Phytopathological models using infected plants (biotic stress) and
4. Field trials

Models Based on Healthy Plants, Microorganisms and Viruses

Among the numerous plant model systems studied, the classical test of wheat germination and growth (Kolisko, 1923) has been repeatedly used as a basic model for research in homeopathic potencies. But also many other organisms were introduced. Table 7.1 gives an overview about the publications released after 1984. Species, treatment, working variable and effect are reported.

Table 7.1 Overview about of bioassays with healthy plants, microorganisms, or viruses: references, experimental main features and observed effect are reported

Reference	Species	Treatment	Working variable	Effect
Andrade, 2001	Chambà•	Homeopathic treatments	Coumarins content	+
Baker and Smith, 1985	Yeast	*Reanalysis of data from Steffen, 1984*	*Periodicity with potencies*	+/–/0
Baumgartner et al., 2004	Dwarf pea	DK, *DH* plant hormones, PÇ	Shoot growth	+/0
Betti et al., 1994	Wheat	*DH* As_20_3	Germination	+
Bonato and da Silva, 2003	Radish	CH, MCH *Sulfur*	Shoot and root growth	+
Bornoroni, 1991	Oat	IAA, CH $CaCO_3$, PC	Shoot growth	+/0
Brack et al., 2003	Bacterium	*DH* 3,5-dichlorophenol	Luminescence	–/0
Endler and Pongratz, 1991	African violet	*DH* IBA, PC	Root and leaf growth	+
Engstler, 2004	Yeast	CH homeopathic treatments***	*In vitro* growth (optical density)	+/–/0
Glatthaar-Saalmüller et al., 2001	Viruses	*D Euphorbium resinifera, Pulsatilla pratensis, Luffa operculata; Euphorbium compositum SN*	Growth of virus plaques	–/0
Grange et al., 1987	Bacteria	CH homeopathic treatments*; mother tinctures**	*In vitro* growth	–/0
Hagelberg, 1987	Yeast	CH homeopathic treatments**, PC	*In vitro* growth (optical density)	0
Hamman et al., 2003	Barley	CH gibberellic acid	Germination, root and shoot growth	+/0
Khanna et al., 1992	Fungi	DH and CH homeopathic treatments*****	Respiration of germinating spores	–/0
KoCH et al., 1995	Yeast	*DH* salicylic acid, $CuSO_4$, $NaNO_2$	*In vitro* growth (no. of cells)	0
Malarczyk et al., 2003	Fungi	CH guaiacol, ethanol	Enzymatic activity	+/–/0
Pongratz and Endler, 1994	Wheat	*DH* $AgNO_3$, PC	Germination, shoot growth	+
Pongratz et al., 1998	Wheat	*DH* $AgNO_3$	Shoot growth	+
Scherr et al., 2007	Duckweed	*DH* homeopathic treatments******, PÇ	Frond (leaf) growth	+/–/0

(continued)

Table 7.1 (continued)

Reference	Species	Treatment	Working variable	Effect
Scherr et al., 2006	Yeasts	*DH* homeopathic treatments****, PÇ	Growth kinetics	–/0
Steffen, 1984	Yeast	CH *Pulsatilla*	*In vitro* growth (no. of cells)	0
Steffen, 1985	Yeast	*DH* $AgNO_3$, $CuSO_4$, $HgCl_2$, NaCl	*In vitro* growth (no. of cells)	0
Tschulakow et al., 2005	Dinoflagellate	Succussed and unsuccussed medium	Bioluminescence	+

D, C, Mc = decimal, centesimal, hundred thousand potency; H = hahnemannian potency; K = korsakovian potency; PC = potentized control (as additional control); IAA = indole acetic acid; IBA = indole butric acid. • = *Justitia pectoralis;**= Belladonna, Drosera, Lycopodium, Plumbago, Prunus spinosa, Pulsatilla nigricans, Pyrethrum parthenium, Raphanus, Rhus toxicodendron, Usnea barbata; **= 31 different mother tinctures; ***= *Sulphur, Arnica montana, Chamomilla, Bryonia alba, Euphrasia officinalis, Pulsatilla;* ***= screening of 49 substances, replications of *Arnica montana, Aurum metallicum, Berberis vulgaris, Colchicum autumnale, Conium maculatum, Corallum rubrum, Cyclamen purpurascens, Lycopodium clavatum, Medorrhinum, Mercurius solubilis Hahnemanni, Natrium chloratum, Pulsatilla pratensis, Solanum dulcamara, Stannum metallicum, Strychnos nux-vomica;* ****= screening of 14 substances, replications of Azoxystrobin, Phosphorus; *****=*Arsenicum album, AsvaganDH, Blatta orientalis, Filix mas, Kali iodatum, Kali muraticum, Lycopodium clavatum, Phosphorus, Thuja occidentalis, Zincum sulphuricum;* ******= screening of 12 substances. + = stimulating or increasing effect; – = inhibiting or decreasing effect; +/– = different effects according to the potency used or physiological conditions; 0 = not significant, i.e. effect below natural variability of the experimental system used.

In particular, Pongratz's results (Pongratz and Endler, 1994; Pongratz et al., 1998) confirmed previous data showing that three consecutive potencies of silver nitrate, a substance which in high concentration inhibits germination, induced a typical 'V'-form effect pattern: 24 and 26 *DH* stimulated and 25 *DH* weakened stalk growth. The simplicity of the model made it possible to repeat the experiment in a multi-centre trial, a very important requirement for the validation of high-potency studies.

In experiments on other plant species growth parameters and biochemical responses were evaluated. Between the most recent papers, three studies have examined the effects of homeopathically prepared plant hormones (e.g. gibberellic acid or kinetin) on the germination performance of barley seeds (Hamman et al., 2003), length growth of dwarf peas (Baumgartner et al., 2004) and frond (leaf-like structures) area growth rate of duckweed (Scherr et al., 2007). In all cases, significant effects have been observed, supporting the hypothesis that homeopathic potencies of plant growth substances may be effective.

Also microorganisms and viruses were used as model systems; several studies have been published in the last years. These included different strains of viruses and species of bacteria, yeasts, fungi and one species of dinoflagellates (see Table 7.1). Though it seems to be easier to use microorganisms (compared to plants) with respect to technical handling, it is nevertheless important to exactly control and document the methodological details of all experimental conditions in order to allow replication experiments. The latter are necessary to verify effects found and to determine all important parameters necessary for successful replication.

Among the studies with microorganisms most often yeasts have been used as model organisms (Steffen, 1984, 1985; Baker and Smith, 1985; Hagelberg, 1987; Koch and Partilla, 1995; Engstler, 2004; Scherr et al., 2006). Some of these works were inspired by the positive results of Jones et al. (Jones and Jenkins, 1983a,b) indicating sensitive reactions of *Saccharomyces cerevisiae* to potencies of *Pulsatilla*. However, also with microorganisms there seem to be some difficulties when external replication of reported results was intended (Steffen, 1984). A re-analysis of the data by Baker (Baker et al., 1985) showed similar periodicities with potencies in the work of Steffen et al. (Steffen, 1984) and Jones et al. (Jones and Jenkins, 1983a, b). In general, the yeast model system was reported to be stable and reliable, however varying sensitivity to homeopathic potencies has been found, depending on the measured parameter and on the substances and potency levels tested (Baker et al., 1985; Engstler, 2004; Scherr et al., 2006). It can be hypothesized that different strains as well as different physiological conditions of the cells at the beginning of the experiment may be important when using microorganisms for investigating effects of potentized substances. In the studies named two different yeast species have been used, either *Saccharomyces cerevisiae* (Steffen, 1984; Koch et al., 1995; Engstler, 2004) or *Schizosaccharomyces pombe* (Steffen, 1985; Hagelberg, 1987). In one study both species were tested and showed differential reactions in their growth kinetics when treated with homeopathic potencies (Scherr et al., 2006). Engstler (Engstler, 2004) did not come to consistent conclusions because of problems with internal repeatability and sterility.

Substance specifity, i.e. the effect that some potentized substances did affect the measured parameters, whilst others did not when tested in the same experimental set-up, was observed for viruses (Glatthaar-Saalmüller et al., 2001), yeast (Scherr et al., 2006) and fungi (Khanna and Chandra, 1992). Another phenomenon was demonstrated in those studies in which several potency levels were tested simultaneously: there were active and inactive potency levels. This phenomenon has also been demonstrated with bacteria (Brack et al., 2003), yeast (Scherr et al., 2006) and fungi (Malarczyk et al., 2003).

The importance of the succussion step in the preparation process of homeopathic remedies was investigated by Tschulakow et al. (2005). They studied the effect of succussed and unsuccussed medium when measuring the intensity of bioluminescence in a dinoflagellate. The differences found were highly significant and independent of the number of succussions (in the range between 13 and 64).

Models Based on Impaired Plants and Microorganisms

In Table 7.2 we summarize recent investigations of homeopathic dilutions in systems with abiotic stress. Species, stress factor, treatment, working variable and effect are reported.

Table 7.2 Overview about of bioassays with abiotic stress: references, experimental main features and observed effect are reported

Reference	Species	Stress	Treatment	Working variable	Effect
Auquière et al., 1988	Wheat	Ethanol, Lysine	CH Ethanol, Lysine, PC	Shoot growth, weight of shoots	+, –
Betti et al., 1997	Wheat	As_2O_3	*DH* As_2O_3, PC	Shoot growth	+
Binder et al., 2005	Wheat	As_2O_3	*DH* As_2O_3, PC	Shoot growth	–
Brizzi et al., 2000	Wheat	As_2O_3	*DH* As_2O_3, PC	Germination	+/–
Brizzi et al., 2005	Wheat	As_2O_3	*DH* As_2O_3, PC	Shoot growth	+
Carvalho et al., 2003	Feverfew	Adaptation	*DH Arnica montana*	Shoot growth, parthenolide content	+, –
Carvalho et al., 2004	Feverfew	Water shortage	CH *Natrum muriaticum*, nosode	Shoot growth, chlorophyll and proline content	0,+/–
Carvalho et al., 2005	Feverfew	Adaptation	CH *Arnica montana*	Shoot growth, parthenolide content	0, –
Egger, 1992	Angels's trumpet	Gamma-irradiation	CH Gamma-radiation, PC	Germination	+, 0
Kovac et al., 1991	Wheat	NaCl, CuCl, $K_2Cr_2O_7$	*DH* NaCl (isopathic), *DH* CuCl (isopathic), *DH* $K_2Cr_2O_7$ (isopathic)	Shoot growth; fresh and dry weight of shoots, grains and roots	+/–/0
Lauppert, 1995	Wheat	Dark germination	*DH* $CuSO_4$	Shoot growth; fresh of shoots; dry weight of shoots, grains and roots	+/–/0
Lehner et al., 1991	Wheat	Dark germination	*DH Platinum, Mercurius, Cadmium, Plumbum, Cuprum, Aurum, Argentum nitricum, Sulfur*	Shoot growth; fresh and dry weight of shoots, grains and roots	+/–/0
Novic et al., 1990	Wheat	Dark germination	*DH Aurum*	Shoot growth; fresh and dry weight of shoots	+/–/0
Projetti et al., 1985	Lentil	$CuSO_4$	CH $CuSO_4$	Root growth	+, 0
Steffen, 1985	Yeast	$CuSO_4$	*DH* $CuSO_4$	*In vitro* growth	0
Tighe, 2005	Cress	NaCl	CH NaCl, PC	Shoot growth, germination	+, 0

D, C = decimal, centesimal potency; H = hahnemannian potency; K = korsakovian potency; PC = potentized control (as additional control); + = stimulating or increasing effect; – = inhibiting or decreasing effect; +/– = different effects according to the potency used or plant physiological conditions; 0 = not significant, i.e. effect below natural variability of the experimental system used.

The work of Bologna University research group (Betti et al., 1994, 1997; Brizzi et al., 2000, 2005) focuses on the statistical analyses of a series of experiments on the same biological model, where a large number of wheat seeds were treated with decimal potencies of arsenic trioxide. The consistency of the different statistical analyses, as well as the reproducibility of most of the experimental results is notable: the As_2O_3 45 *DH* potency always induces a highly significant stimulating effect compared to control, as well as H_2O at the same potency, whereas As_2O_3 diluted at 10^{-45} never show any effect. The reported results confirm that the potentization process is critical to make raise the biological effects secondary to the treatments with respect to control.

Moreover, wheat germination is the theme jointly investigated by Betti and Baumgartner research groups: the result of Baumgartner replication trial (Binder et al., 2005) is a reversal of the original study (Betti et al., 1997), since *Arsenicum album* 45 *DH* inhibited wheat shoot growth instead of enhancing it, whereas Betti replication trial (Brizzi et al., 2005) reassessed the result of its initial study (Betti et al., 1997). Nevertheless, high homeopathic potencies induced statistically significant effects in both experiments, even if the magnitude and direction of these effects seem to depend on yet unknown parameters (Binder et al., 2005). Elucidation of factors responsible for effect size and direction will yield important information, possibly helping to understand the reasons for problems with reproducibility also in other systems.

In some studies it is not easy to decide, if stressed or healthy organisms were used, because in some cases there was no active intervention to cause damage, and in other cases the damages were not exactly defined. For example, Novic et al. investigated the effect of potentized gold on shoot growth of wheat seeds. To increase the sensitivity of the system they aimed to create deficiency conditions through germination in darkness and without addition of nutrients (Novic et al., 1990). Lehner et al. used the same experimental design to test metals like *Platinum*, *Mercurius*, *Cadmium*, *Plumbum*, *Cuprum*, *Aurum* as well as *Argentum nitricum* and *Sulfur* under mentioned deficiency conditions (Lehner et al., 1991). Dark germination is a natural condition for wheat, however. To what extent germinating seeds were stressed by unavailability of nutrients, was not specified. Lauppert used the same experimental design, accounting that germination of seeds in darkness was chosen only to get uniform conditions, but not with the aim of stressing the organisms (Lauppert, 1995).

Carvalho et al. investigated the effect of low potentized *Arnica montana* on stressed feverfew caused by adaptation (Carvalho et al., 2003, 2005). In one study they used *Natrium muriaticum* for the treatment of damages induced by water shortage (Carvalho et al., 2004). Kovac et al. worked with wheat seeds germinated over 6 days in darkness without nutrient and, in addition, supplied with high NaCl-concentrations as abiotic stress factor (Kovac et al., 1991). They emphasized the need of a reduction of variance (see above) to achieve clear results. Tighe also used NaCl for weakening of plants. He chose cress as test organism after comparing wheat, sunflower and cress in preliminary tests (Tighe, 2005). Egger treated *Datura arborea* seeds with gamma radiation and treated the plants with potentized gamma radiation (gamma irradiated lactose; Egger, 1992).

It is needed to add some methodological considerations on the use of impaired or stressed organisms in preclinical model systems.

All experiments with poisoned plants (as listed in Table 7.2, except one preliminary test from Egger (1992)), used the isopathic approach. Thus, the problem of finding an appropriate remedy according to the *similia principle* could be "avoided". A further advantage of this experimental approach is the possibility to test true *information* effects even in low potencies (low potencies of many substances can influence plants due to molecular – non homeopathic effects; given a pre-existing damage by higher concentrations of the same substance, any effect of a treatment with lower concentration cannot be explained by the material presence of this substance). In addition, low isopathic doses can be a useful tool to validate or not some principles of anthroposophical medicine, in which low potencies might act on a *deeper organizational* level of the organism. Plant models are especially interesting for studying this matter.

In addition, it might be promising in future to try other approaches than the isopathic one in systems with poisoned organisms. Ways to approximate classical homeopathy may be the use of phenomenological or biochemical symptoms on basis of the *similia principle.*

If a systematic reduction of variability is – as aforementioned – a specific effect of homeopathic potencies, i.e. stronger in situations that allow a regulative response of the organism, the homeopathic/isopathic treatment of organisms disturbed by stress should exceed the effect of treatments of healthy organisms in basically equalized conditions. Furthermore, reproducibility might be enhanced if the damaging effect and the corresponding specific organic answer to the homeopathic/isopathic treatment are stronger than the unspecific noise concomitant factors.

The variances of model systems under abiotic stress are highly dependent on the concentration of noxae used. A straight relation has to be found between a measurable damage and a sufficient fitness of the organism that still allows self-healing. The variance increases strongly up to a certain concentration of noxae, thereby causing a very instable system. The further reduction of the organism vital functions leads to a lower system-sensitivity and consequent decreasing of variance, in a second step. Hence, variance and sensitivity correlate. For this reason, a sensitive system with weak organisms has to be highly stabilized to prevent loss of statistical power caused by escalating variance. Anyway, it is important to remember that the equilibrating effect of homeopathic preparations does not necessarily imply a reduction of variance in any case.

Phytopathological Models

In most of the papers available focused on fungal infections (Saxena et al., 1987; Khanna and Chandra, 1989, 1992; Aggarval et al., 1993; Rivas et al., 1996; Rolim et al., 2001) following homeopathic treatments, decrease of disease symptoms, post-harvest losses, fungal germination and respiration rate of germinating spores were evidenced. Few studies took into account viral infections (Cheema, 1986, 1993; Betti et al., 2003b) and, in this case, a weaker symptomatology was observed. In particular, in blind randomized experiments using tobacco plants carrying tobacco mosaic virus (TMV) resistant gene *N*, a significant enhancement of plant resistance was obtained following As_2O_3 5 and 45 *DH* potencies (Betti et al., 2003b). As far as nematode infection is concerned, a few papers are available as well (Sukul and Sukul, 1999, 2006; Datta, 2006): plants treated with homeopathic preparations showed improved growth (in terms of shoot and root length) and reduced nematode infection (in terms of root gall number and nematode population in root and soil). Root- and leaf-protein content and root-water content were also affected by homeopathic treatments. A brief review about is seen in Table 7.3.

Table 7.3 Summary of phytopathological bioassays: references, experimental main features and observed effect are reported

Reference	Species/pathogen	Treatment	Working variable	Effect
Aggarval et al., 1993	Wild taro/ *phytophthora colocasiae*	Homeopathic treatments**	Disease symptoms, fungal growth and spore germination	–
Betti et al., 2003b	Tobacco/*tobacco mosaic virus*	*DH* As_2O_3, PC	Virus-induced hypersensitive lesions	–
Cheema et al., 1986	Papaya/*papaya mosaic virus*	Homeopathic treatments	Disease symptoms	–
Cheema et al., 1993	Tomato/*tobacco mosaic virus*	*Clerodendrum aculeatum*, CH *Thuja*	Disease symptoms	–
Datta, 2006	Mulberry/ *m. Incognita*	CH *Cina*	Plant growth, nematode infection	+, –
Khanna and Chandra, 1989	Mango, guava, tomato/ *pestalotia spp., fusarium roseum*	Homeopathic treatments and adjuvants	Post-harvest losses	–
Khanna and Chandra, 1992	Different fungi••	*DH* treatments	Spore respiration rate, organic acid pool in spores	– +/–
Rivas et al., 1996	Wheat, tomato/ *Alternaria solani*	CH treatments***	Seed and spore germination	+/–
Rolim et al., 2001	Apple/ *podosphaera leucotricha*	CH treatments****	Powdery mildew symptoms	–
Saxena et al., 1987	Reed okra/ seed-borne fungi	CH *Thuja*, nitric acid, *Sulphur, Calcarea carb., Teucrium Q*	Fungal spore germination	–

(continued)

Table 7.3 (continued)

Reference	Species/pathogen	Treatment	Working variable	Effect
Sukul and Sukul, 1999	Cowpea / *Meloidogyne incognita*	CH *Cina*	Plant growth, nematode infection	+ –
Sukul et al., 2006	Lady's finger/ *M. incognita*	CH *Cina, Santonin, Ethanol*	Nematode infection root-protein and –water content	– –

D, C = decimal, centesimal potency; H = hahnemannian potency; PC = potentized control (as additional control); •• = *Alternaria alternata, Colletotrichum gloeosporioides, Fusarium roseum, Gloeosporium psidii, Pestalotia mangiferae, Pestalotia psidii;* ** =*Kali iodatum, Arsenicum album, Blatta orientalis, Thuja occidentalis;* ***= *Arsenicum album, Calcarea, Cuprum, Ferrum metallicum, Lycopodium, Natrum, Phosphorus, Selenium, Silicea, Sulphur;* **** = *Kali iodatum, Lachesis trigonocephalus, Staphysagria, Sulphur, Oidium lycopersici.* + = stimulating or increasing effect; – = inhibiting or decreasing effect; +/– = different effects according to the potency used or plant physiological conditions.

Field Trials

Scientific literature provides very few and outdated descriptions of field trials (Table 7.4). Aside from two studies on trees affected by virus (Sinha, 1976) or fungus (McIvor, 1980), only two papers are easily available. Kayne reported the results of a field trial on rye grass (Kayne, 1991). The application of homeopathic sprays (CH *Sulphur* and mixture of CH *Sulphur*, *Silicea* and *Carbo vegetalis*) did not give significant effects on plant growth; however some methodological hints for testing homeopathic treatments emerged: the choice of remedy, potency and frequency of application is crucial and should be made with great care to ensure the best chance of success. In a more recent paper (Diniz, 2006) the efficacy of a homeopathic preparation for tomato late blight control was evaluated. The treatment was prepared from tomato tissue infected by *Phytophthora infestans*, agent of tomato late blight (isopathic treatment at 30 CH potency). Also in this case, no significant effect has been observed with respect to control.

A three year project (2004–2006) on biological control of dark leaf spot caused by *Alternaria brassicicola* in cauliflower was financed by the Marche region (Italy) to the Betti research group. The field trial results showed that As_2O_3 *DH* 35 could significantly reduce infection level on cauliflower heads with respect to control. A resistance increase in tobacco plants against tobacco mosaic virus following treatments with As_2O_3 45 *DH* has been already reported (Betti et al., 2003b) and here it was shown the significant effects of As_2O_3 35 *DH* in the control of dark leaf spot disease. It is noteworthy that in different plant/pathogen interactions different potencies of the same substance have different efficacy. Since As_2O_3 35 *DH* was diluted above Avogadro's number, there were no arsenic molecules in the treatment, thus, it can be used in agricultural practice without introducing pondered arsenic into the environment. These results need further investigations, but they seem to support the

Table 7.4 Summary of field trials: references, experimental main features and observed effect are reported

Reference	Species/pathogen	Treatment	Working variable	Effect
Diniz et al., 2006	Tomato/ *Phytophthora infestans*	CH isopathic treatment	Late blight symptoms	0
Kayne, 1991	Rye grass	CH *Sulphur, Silicea, Carbo vegetalis*	Plant growth	0
Trebbi et al., 2008	Cabbage/ *Alternaria brassicicola*	*DH* As_2O_3	Disease severity	–

D, C = decimal, centesimal potency; H = hahnemannian potency; – = inhibiting or decreasing effect; 0 = not significant, i.e. effect below natural variability of the experimental system used.

possibility of an agricultural application of potentized drugs. The privileged target of agro homeopathy could be small farms (and in particular, those of nutraceutical and herbalist sectors) practicing organic and sustainable agriculture that strive to be environmentally responsible, economically viable, and socially just.

However, putative adverse effects for cattle and consumer health must also be considered. Remains of the potentized remedy itself, as well as processes induced in plants, might cause reactions in animals and humans. Given the fact that the exact nature of these preparations is still unknown and that there is not any analytical methods to detect application of homoeopathic/isopathic potencies, it would be advisable to proceed with caution. Application of homoeopathic/isopathic potencies in organic plant production should be tested thoroughly before any products are sold to the public.

Conclusions and Perspectives

The literature on homeopathy or isopathy and plants or microorganisms is limited and not always easily available. Nevertheless, interest in this field appears to be growing in recent years and several projects are in progress, mainly in Central and South America. In general, the potential prospects for such treatments in plant and microbial basic research and agriculture can be considered promising. In particular, much more work is needed, especially at field level, since the influence of environmental and agronomical factors (temperature, drought, humidity, plant cultivars and so on) might significantly change the quality of yields and, thus, the results of successive experimentations. In addition, there is no certain knowledge about the most effective potency level or potency type (d/c), and neither adequate doses nor application frequencies of potentized dilutions for plants are sufficiently known.

The use of plants and microorganisms in homeopathic basic research has a considerable potential. We think that it is possible to use such model systems to elucidate basic nature and working principles of homeopathic/isopathic preparations and to develop quality control instruments (e.g. for optimization of production, storage, and transport procedures).

Finally, it has to be stressed that results of all research and projects, whether successful or not, should be made widely available so that others can learn from these, avoiding duplication and inefficiency. Moreover, replication of results and multicentre trials should be performed, to be published in international journals with an impact factor or wide circulation, to gain credibility and facilitate funding.

References

Aggarval A, Kalmesh T, Mehrotra R S. Control of Taro Blight and Corn Rot Caused by Phytophtora colocasiae Homeopathic Drugs. Plant Disease Research; 1993, 8: 94–101.

Andrade F M C. Efeito De Homeopatias No Crescimento E Na Produção De Cumarina Em Chambà (*Justicia Pectoralis* Jacq). Rev Bras Plantas Med; 2001, 4: 19–28.

Auquière J P, Moens P, Linster T, Speileux M. Recherche de l'action de dilutions homéopathiques sur les végétaux 3. Essais de deux substances organiques métabolisables: L'etanol et la lysine sur le blé. In: Kritische Begutachtung der wissenschaftlichen Beweisführung in der homöopathischen Grundlagenforschung. Hannover: Majerus, M. Hannover; 1988.

Baker R D, Smith C W. Comment on the Paper, Growth of Yeast Cultures as in Vitro Model for Investigating Homoeopathic Remedies. The British Homoeopathic Journal; 1985, 74(2): 93–96.

Baumgartner S, Heusser P, Thurneysen A. Methodological Standards and Problems in Preclinical Homoeopathic Potency Research. Forschende Komplementärmedizin; 1998, 5: 27–32.

Baumgartner S, Thurneysen A, Heusser P. Growth Stimulation of Dwarf Peas (*Pisum sativum* L.) through Homeopathic Potencies of Plant Growth Substances. Forschende Komplementärmedizin und Klassische Naturheilkunde; 2004, 11: 281–292.

Betti L, Brizzi M, Nani D, Peruzzi M. A Pilot Statistical Study with Homoeopathic Potencies of *Arsenicum Album* in Wheat Germination as a Simple Model. British Homoeopathic Journal; 1994, 83(4): 195–201.

Betti L, Brizzi M, Nani D, Peruzzi M. Effect of High Dilutions of *Arsenicum Album* on Wheat Seedlings from Seed Poisoned with the Same Substance. British Homoeopathic Journal; 1997, 86(April): 86–89.

Betti L, Borghini F, Nani D. Plant Models for Fundamental Research in Homeopathy. Homeopathy; 2003a, 92: 129–130.

Betti L, Lazzarato L, Trebbi G, Brizzi M, Calzoni G L, Borghini F, Nani D. Effects of Homeopathic Arsenic on Tobacco Plant Resistance to Tobacco Mosaic Virus. Theoretical Suggestions About System Variability, Based on a Large Experimental Data Set. Homeopathy; 2003b, 92: 195–202.

Betti L, Lazzarato L, Trebbi G, Nani D. The Potential and Need for Homeopathy Research in Horticulture and Agriculture. Improving the Success of Homeopathy: A Global Perspective. 2006; London.

Binder M, Baumgartner S, Thurneysen A. The Effects of a 45x Potency of *Arsenicum Album* on Wheat Seedling Growth – a Reproduction Trial. Forschende Komplementärmedizin und Klassische Naturheilkunde; 2005, 12(5): 284–291.

Bonato C M, da Silva E P. Effect of the Homeopathic Solution Sulfur on the Growth and Productivity of Radish. Acta Scientiarum. Agronomy; 2003, 25: 259–263.

Bornoroni C. Synergism of Action between Indoleacetic Acid (IAA) and Highly Diluted Solutions of $CaCO_3$ on the Growth of Oat Coleoptiles. The Berlin Journal on Research in Homoeoepathy; 1991, 1(4/5): 275–278.

Brack A, Strube J, Stolz P, Decker H. Effects of Ultrahigh Dilutions of 3,5-Dichlorophenol on the Luminescence of the Bacterium *Vibrio fischeri*. Biochimica et Biophysica Acta; 2003, 1621: 253–260.

Brizzi M, Nani D, Peruzzi M, Betti L. Statistical Analysis of the Effect of High Dilutions of Arsenic in Large Dataset from a Wheat Germination Model. British Homeopathic Journal; 2000, 89: 63–67.
Brizzi M, Lazzarato L, Nani D, Borghini F, Peruzzi M, Betti L. A Biostatistical Insight into the As_2O_3 High Dilution Effects on the Rate and Variability of Wheat Seedling Growth. Forsch Komplementarmed Klass Naturheilkd; 2005, 12(5): 277–283.
Carrington J C, Ambros V. Role of MicroRNAs in Plant and Animal Development. Science; 2003, 301: 336–338.
Carvalho L M, Casali V W D, Cecon P R, Souza M A, Lisboa S P. Effect of Decimal Potencies of Homeopathy *Arnica Montana* on Plants of Feverfew. Rev Bras Plantas Med; 2003, 6: 46–50.
Carvalho L M, Casali V W D, Cecon P R, Lisboa S P, Souza M A. Effect of Homeopathy on Recovery of Feverfew Plants, *Tanacetum partenium* (L.) Schultz Bip, under Water Stress. Rev Bras Plantas Med; 2004, 6: 20–27.
Carvalho L M, Casali V W D, Lisboa S P, Souza M A, Cecon P R. Effect of Homeopathy *Arnica Montana*, Centesimal Pontencies, on Plants of Artemisia. Rev Bras Plantas Med; 2005, 7(3): 33–36.
Cheema S S, Reddy R S, Kapur S P, Bansal R D. Comparative Comparative Efficacy of Homeopathic Drugs against Papaya Mosaic Virus (PapMV) as Foliar Spray. Indian Journal of Virology; 1986, 2: 132–135.
Cheema S S, Sanjiv-Kapila, Arun-Kumar, Kapila S, Kumar A. Efficacy of Various Bio-Products and Chemical against Tobacco Mosaic Virus in Tomato and Cucumber Mosaic Virus in Bottle Gourd. Plant Disease Research; 1993, 8: 110–114.
Datta S C. Effects of *Cina* on Root-Knot Disease of Mulberry. Homeopathy; 2006, 95: 98–102.
Diniz L P, Maffia L A, Dhingra O D, Casali V W D, Santos R H S, Mizubuti E S G. Avaliação De Produtos Alternativos Para Controle Da Requeima Do Tomateiro. Fit Bras; 2006, 31: 171–179.
Egger M. The Effect of Homeopathic High Potencies on Gamma-Irradiated *Datura Arborea* Seeds. Zumikon, Schweiz, Antioch University; 1992.
Endler P C, Pongratz W. Homeopathic Effect of a Plant Hormone? A Preliminary Report. The Berlin Journal on Research in Homeopathy; 1991, 1: 148–150.
Engstler G. Influence of High Diluted Substances on the Growth of Genetically Identical Yeast Strains. Thesis in Medizinische Fakultät. Wien, Medizinische Universität Wien; 2004.
Glatthaar-Saalmüller B, Fallier-Becker P. Antiviral Action of *Euphorbium compositum*® and Its Components. Forschende Komplementärmedizin und Klassische Naturheilkunde; 2001, 8: 207–212.
Grange J M, Davey R W, Jonas S K. A Study of the Bacteriocidal and Bacteriostatic Effects of Preparations Derived from Plant Material Used in Herbal and Homoeopathic Medicine. Complementary Medical Research; 1987, 2(2): 135–140.
Hagelberg E. The Effect of Homoeopathic Medicines on Yeast Growth. The British Homoeopathic Journal; 1987, 76(3): 126–129.
Hamman B, Koning G, Him Lok K. Homeopathically Prepared Gibberellic Acid and Barey Seed Germination Performance. Homeopathy; 2003, 92: 140–144.
Hyland M E, Lewith G T. Oscillatory Effects in a Homeopathic Clinical Trial: An Explanation Using Complexity Theory, and Implications for Clinical Practice. Homeopathy; 2002, 91(3): 145–149.
Jones R L, Jenkins M D. Comparison of Wheat and Yeast as *In Vitro* Models for Investigating Homoeopathic Medicines. The British Homoeopathic Journal; 1983a, 72(3): 143–147.
Jones R L, Jenkins M D. Effects of Hand and Machine Succussion on *In Vitro* Activity of Potencies of *Pulsatilla*. The British Homoeopathic Journal; 1983b, 72(4): 217–223.
Junker H. Die Wirkung extremer Potenzverdünnungen auf Organismen. Pflügers Archiv für die Gesamte Physiologie; 1928, 219 B(5/6): 647–672.
Kayne S. An Agricultural Application of Homeopathy. The British Homoeopathic Journal; 1991, 80: 157–160.
Khanna K K, Chandra S. Further Investigations on the Control of Storage Rot of Mango, Guava and Tomato Fruits with Homeopathic Drugs. Indian Phytopathology; 1989, 42: 436–440.

Khanna K K, Chandra S. Effect of Homoeopathic Drugs on Respiration of Germinating Fungal Spores. Indian Phytopathology; 1992, 45(3): 348–353.

Koch H P, Partilla A. Experimentelle Untersuchungen zur Wirksamkeit homöopathischer Potenzen im Hefe-Test. Pharmazie; 1995, 50(1): 73.

Kolisko L. Physiologischer und physikalischer Nachweis der Wirksamkeit kleinster Entitäten. Stuttgart, Der Kommende Tag AG; 1923.

Kovac H, Muhry F, Novic S, Moser M. Das Wachstum von Weizenkeimlingen nach Zugabe von toxischen Substanzen. Mitteilungen des Instituts für Strukturelle Medizinische Forschung; 1991, 3: 43–63.

Lam E, Kato N, Lawton M. Programmed Cell Death, Mitochondria and the Plant Hypersensitive Response. Nature; 2001, 411(6839): 848–853.

Lauppert E. Auswirkung 'homöopathisch' zubereitetem Kupfersulfat auf das Wachstum von Weizenkeimlingen. Thesis in Institut für Pflanzenphysiologie. Karl-Franzens-Universität Graz: 115; 1995.

Lehner E, Muhry F, Kovac H, Novic S, Pongratz W, Moser M. Untersuchung der Wirkung von potenzierten Umweltschadstoffen (Metallen) und Schwefel auf das Wachstum von Weizenkeimlingen. Mitteilungen des Instituts für Strukturelle Medizinische Forschung; 1991, 3: 1–42.

Malarczyk E, Jarosz-Wilkolazka A, Kochmanska-Rdest J. Effect of Low Doses of Guaiacol and Ethanol on Enzymatic Activity of Fungal Cultures. Nonlinearity in Biology, Toxicology and Medicine; 2003, 1(2): 167–178.

McIvor G. Letters to Editor. Journal of the American Institute of Homeopathy; 1980, 73: 43–48.

Nani D, Brizzi M, Lazzarato L, Betti L. The Role of Variability in Evaluating Ultra High Dilution Effects: Considerations Based on Plant Model Experiments. Forschende Komplementärmedizin; 2007, 14(5): 301–305.

Novic S, Muhry F, Lehner E, Kovac H, Pongratz W, Grivetz S, Moser M. Untersuchung der Wirkung von potenziertem Gold (*Aurum Met.* Praep.) auf das Wachstum von Weizenkeimlingen. Mitteilungen des Instituts für Strukturelle Medizinische Forschung; 1990, 2: 14–43.

Pongratz W, Endler P C. Reppraisal of a Classical Botanical Experiment in Ultra High Dilution Research. Energetic Coupling in a Wheat Model. In: Endler PC, Schulte J (Eds.) Ultra High Dilution. Physiology and Physics. Dordrecht: Kluwer; 1994.

Pongratz W, Nograsek A, Endler P C. Highly Diluted Agitated Silver Nitrate and Wheat Seedling Development. Effect Kinetics of a Process of Successive Agitation Phases. In: Schulte J, Endler PC (Eds.). Fundamental Research in Ultra High Dilution and Homeopathy. Dordrecht: Kluwer; 1998.

Projetti M L, Guillemain J, Tetau M. Effects curatifs et préventifs de dilutions homéopathiques de sulfate de cuivre appliquées à des racines de lentilles pré- ou post-intoxiquées. Cahiers de Biotherapie; 1985, 88: 21–27.

Rivas E, Ceceña C, Guajardo G. Acción De 9 Fármacos Homeopáticos Sobre La Germinacion De Esporas De *Alternaria Solani* Y Semillas De Trigo Y Tomate. Boletin Mexicano de Homeopatia; 1996, 29: 44–46.

Rolim P R R, Brignani Neto F, Silva J M. Controle de oídio da Macieira por preparações homeopáticas. Fit Bras; 2001, 26(supl. agosto): 435–436.

Saxena A, Pandey M L, Gupta R C. Effect of Certain Homeopathic Drugs on Incidence of Seed-Borne Fungi and Seed Germination of *Abelmoschus Esculentus*. Indian Journal of Mycology and Plant Pathology; 1987, 17: 191–192.

Scherr C, Baumgartner S, Spranger J, Simon M. Effects of *Potentized* Substances on Growth Kinetics of *Saccharomyces cerevisiae* and *Schizosaccharomyces pombe*. Forschende Komplementärmedizin; 2006, 13: 298–306.

Scherr C, Simon M, Spranger J, Baumgartner S. Duckweed (*Lemna gibba* L.) as a Test Organism for Homeopathic Potencies. Journal of Alternative and Complementary Medicine; 2007, 13(9): 931–937.

Scofield A. Homeopathy and Its Potential Role in Agriculture, a Critical Review. Biological Agriculture and Horticulture; 1984, 2: 1–50.

Sinha E. Agro-Homeopathy. Journal of the American Institute of Homeopathy; 1976, 68: 37–40.
Steffen W A. Growth of Yeast Cultures as *In Vitro* Model for Investigating Homoeopathic Medicines – a Critical Assessment. The British Homoeopathic Journal; 1984, 73(4): 198–210.
Steffen W A. Growth of Yeast Cultures as *In Vitro* Model for Investigating Homoeopathic Medicines – Some Further Studies. The British Homoeopathic Journal; 1985, 74(3): 132–140.
Sukul N C, Sukul A. Potentized *Cina* Reduced Root-Knot Desease of Cowpeas. Enviroment and Ecology; 1999, 17(2): 269–273.
Sukul N C, Ghosh S, Sukul A, Sinhababu S P. Amelioration of Root-Knot Disease of Lady's Finger Plants by Potentized *Cina* and *Santonin*. Homeopathy; 2006, 95(3): 144–147.
Tighe M. Homeopathy on Cress – a Pre-Clinical Method Development Using a Simple Plant Growth Model. Thesis. University of Westminster: London; 2005.
Trebbi G, Fantino M G, Dinelli G, Marotti I, Burgio G, Nani D, Iero D, Betti L. Effects of Homeopathic and Mineral Treatments on Dark Leaf Spot Caused by *Alternaria brassicicola* on Cauliflower. 16th IFOAM Organic World Congress, Modena, Italy; 2008.
Tschulakow A V, Yan Y, Klimek W. A New Approach to the Memory of Water. Homeopathy; 2005, 94: 241–247.

Chapter 8
Arnica montana and Behavior of Connective Tissue

Leoni Villano Bonamin[1,2]

Introduction

The effects of *Arnica montana* extracts are known since the middle age (Carvalho, 2000). Due to the wide variety of active principles, vesicular inflammatory cutaneous reaction and analgesic, anti-ulcerative, anti-hypertensive, anti-genotoxic and anti-hemorrhagic effects have been classically described (Genet, 1980; Demarque, 1985; Cerqueira et al., 1987; Chakrabarti, 1991; Robles et al., 1995; Rigamonti, 1995; Chakrabarti et al., 2001; Jeffrey and Belcher, 2002; Macedo et al., 2004). Moreover, its most studied active principle, Helenalin, is the main principle involved in the anti-inflammatory effects (Hall et al., 1980; Schimidt et al., 1993; Bucay, 1995; Lyss et al., 1997; Klass et al., 2002; Macedo et al., 2004). The inflammatory process involves a series of tissue and plasma events, in which several kinds of cells and chemical mediators communicate one to another in a web structured organization. This allows reaching an optimal tissue response to an aggressive stimulus (Cotran et al., 2004). Although the use of *Arnica montana* in homeopathic preparations in the treatment of acute inflammation is quite traditional, and is even recommended as an emergency treatment of polytraumatic injuries (Oberbaum et al., 2003), there are some controversies in the literature about it (Ernst and Pittler, 1998; Ramelet et al., 2000; Wolf et al., 2003; Stevinson et al., 2003; Macedo et al., 2004; Conforti et al., 2007), thus, more specific studies about physiopathological changes after Arnica treatments are needed.

The studies of the effects of *Arnica montana* 6CH in acute inflammation and connective tissue have been performed by our group since 1998, when we questioned about which chemical mediators would be involved in its effects upon acute inflammatory process (Carvalho, 2000). In a first study, we compared the effects of the mother tincture and several centesimal homeopathic diluted forms of *Arnica montana* in relation to acute edema induced by 1% carrageenan subcutaneous inoculation in

[1]Universidade Paulista. São Paulo, Brazil. leonibonamin@gmail.com

[2]Universidade de Santo Amaro. São Paulo, Brazil

Presented at the XXIth Symposium of GIRI, Monaco, 2007

L.V. Bonamin (ed.) *Signals and Images*,

rats (SIGMA). The results showed that the homeopathic form of Arnica, mainly *Arnica montana* 6CH, was more effective to reduce the edema peak. Even more, this reduction of edema was not related to any reduction in the vascular permeability but was related to histamine and prostaglandin mediation (Carvalho and Bonamin, 2001). These initial observations were the motivation to continuing the studies, focusing the modifications in vascular and connective tissue behavior in rats treated with *Arnica montana* 6CH and submitted to a model of acute inflammation.

Methods and Experimental Designs

Experimental Subjects

Male Wistar rats weighing 250–300 g were used in all performed experiments. They were maintained in conventional plastic cages (five rats per cage) and in controlled environmental conditions, it means, temperature set at 25 ± 3°C and light/dark cycle set as 12/12 hours, lights on at 06:00. Food and water were offered *ad libitum*.

Treatment

The rats were randomly divided in four groups:

Group 1 – treated with *Arnica montana* – 5% mother tincture
Group 2 – treated with *Arnica montana* 6CH (prepared in 5% hydro-alcoholic solution)
Group 3 – treated with dexametasone, 1.6 mg/kg
Group 4 – treated with 5% hydro-alcoholic solution

The treatments were done *per os*, each 15 minutes during 2 hours, using an automatic pipette (10 µl/100 g body weight). This treatment protocol was based on the traditional clinical use of Arnica in acute inflammatory situations. The first treatment was done immediately after the carrageenan inoculation. The treatments were done in blind, except for dexamethasone, whose liquid characteristics were too much different from the other preparations. Animals were located in a number of five per cage. Each cage corresponded to one exclusive treatment, to avoid "cross-effects" among animals in the same cage (Conforti et al., 2007).

All Arnica and hydro-alcoholic solutions were prepared by *Aleph Homeopatia e Produtos Naturais Ltda.*, São Paulo, SP, Brazil; a homeopathic pharmacy recommended by ANVISA (Brazilian Sanitary Vigilance Agency). All homeopathic preparations of *Arnica montana* were done according to the Brazilian Homeopathic Pharmacopoeia.

Inflammation Induction

The acute inflammation was induced by two methods: (1) the subcutaneous injection of 0.06 ml of 1% carrageenan *kappa* (SIGMA) into the footpad. In this case, the edema was measured using a micrometer (MYTUTOYO); and (2) the 0.1 ml intra-pleural injection of 1% carrageenan *kappa* (SIGMA), diluted in sterile saline. In this case, animals were sacrificed and pleural washings were done 1, 4 and 6 hours after inoculation, using PBS (2 ml) and heparin (25 UI/ml), in order to harvest and count the inflammatory cells. A modified Neubauer chamber was used to perform the absolute counting, using the Trypan's blue method to identify degenerated and viable cells. Smears fixed in absolute methanol (10 minutes/ 4°C) and stained by hematoxylin/eosin method were performed in order to allow differential cell counting. Five hundred cells were counted per smear. Lymphocytes, monocytes and PMN (polymorphonuclear) cells percentage were determined according to cell morphology.

The calculation of edema volume was obtained from the measurements of paw thickness (axis a) and paw width (axis b), using a micrometer (MYTUTOYO). The elliptical volume was calculated according the formula (Giraldi et al., 1994):

$$\text{Vol} = \pi/6 \text{ x a2 x b}$$

Pain Observation

After the intra-pleural administration, the rat behavior was monitored to verify if there was any pain demonstration. The behavior was qualified according the following scores, established at the moment of sacrifice:

Score 0 – normal locomotion and feeding = no pain
Score 1 – few locomotion and feeding
Score 2 – few locomotion and no feeding at all
Score 3 – no locomotion and no feeding at all

Lymphatic Absorption

Rats were inoculated into the footpad with 1% carrageenan, as described above. Three hours after inoculation, each animal was re-inoculated with 0.08 ml of 2.5% Evans Blue Stain (SIGMA) in the same local. After 20 minutes, animals were sacrificed by guillotine to provide the maximum bleeding and the local lymph node was harvested and put into a sterile tube containing 2 ml of formamide (MERCK), where it was maintained for 48 hours at 37°C, in order to extract the stain. The intensity of formamide color was measured in a spectrophotometer (SPECTRONIC

20D ABALYSER), whose wave length was adjusted to 620 nm. A standard curve was used to calculate the concentration of stain extracted from the lymph nodes. This method was adopted from Macedo et al. (2004) and Moriguchi et al. (2004).

The principle of this method is very simple: Evans blue stain links spontaneously with plasma proteins, mainly albumin. Due to the increase of vascular permeability, a proportional quantity of albumin is flown out to extra vascular space together with water and salts (edema) and is drained by lymphatic vessels. Thus, the injection of Evans blue stain into the footpad subcutaneous tissue during the peak of vascular permeability allows verifying the intensity of lymphatic drainage.

Histopathological Examination

After sacrifice, footpads and lungs were harvested and fixed in 10% buffered formalin to posterior histopathological examination. Hematoxylin – Eosin, Toluidine blue, PAS and Verhoeff staining methods were used, to evaluate different aspects of connective tissue organization following Arnica treatment. The vessels diameters were measured using the software Image Tool 3.0. The differential counting of degranulated and non-degranulated mast cells was done by visual method, in which 100 cells/footpad or lung surface were analyzed.

Statistical Analysis

The Bartlett test was used in all experiments to determine if the samples were in accordance with Gaussian distribution. After that, ANOVA/Tuckey-Krammer were applied for edema and cell counting; ANOVA/Dunnet were applied to lymphatic absorption; Kruskal-Wallis/Dunn were applied for pain scores and histometric vessels measurements and X^2 was used for differential counting of leukocytes and mast cells. The value of $p \leq 0.05$ was considered significant. All statistical analysis were planned *a priori*, except multiple comparisons.

Experimental Design and Results

Experiment 1 – Cell Migration in Pleurisy Model

In this experiment, 1% carrageenan was inoculated into the pleural cavity of rats, as described above, and pleural washing was done 1, 4 and 6 hours after, in order to count inflammatory cells. Eight rats were used by group. The preliminary results of this experiment were related by Bonamin et al., (2000).

Regarding the total leukocytes, only dexamethasone reduced cell migration, which could be seen after 4 hours from carrageenan inoculation. The number of degenerated leukocytes was higher also in dexamethasone treated animals, after 1 hour from inoculation (Table 8.1).

The differential counting revealed that animals treated with *Arnica montana* 6CH presented persistence in PMN cells migration up to 4 hours from irritant inoculation. Also, after 4 hours, a higher number of lymphocytes were also seen in the pleural washing. No differences among the groups were seen after 6 hours (Table 8.1).

Significant reduction of pain scores were seen in animals treated with *Arnica montana* 6CH and dexamethasone 1 hour after inoculation, but statistical significance was obtained only in the last case. After that, in 4 and 6 hours, only dexamethasone treated rats presented mild reduction in pain scores regarding to control (Table 8.2).

After sacrifice, all animals were submitted to necropsy and both lungs were fixed in 1% buffered formalin. The histopathological slides were stained by Toluidine blue method to verify the percentage of degranulated mast cells on pleural surface. No differences were seen among groups in all tested times (Table 8.2).

Table 8.1 Total and differential leukocytes counting, from rats bearing pleurisy induced by carrageenan. Cells were harvested by pleural washing, 1, 4 and 6 hours after carrageenan inoculation. (#) Values represented by mean±standard deviation. (##) Values represented by percentage. *ANOVA/Tuckey-Krammer, p≤0.05 in relation to other groups. ** X^2, p≤0.009 in relation to other groups

Time/ parameter	*Arnica montana* Mother tincture	*Arnica montana* 6CH	Dexamethasone	5% hydro-alcoholic solution
One hour				
Total leukocytes (#)				
Viable	55.77 ± 46.58	91.6 ± 144.21	77.53 ± 112.49	63.38 ± 70.71
Dead	0.66 ± 1.03	0.85 ± 0.89	2.28 ± 1.97*	0.85 ± 1.86
PMN cells (##)	91.3	92.1	91.2	90.0
Monocytes	8.2	7.5	8.4	9.6
Lymphocytes	0.5	0.4	0.4	0.4
Four hours				
Total leukocytes				
Viable	33.57 ± 19.78	30.85 ± 0.22	15.42 ± 6.75*	24.85 ± 7.94
Dead	0.57 ± 0.53	0.71 ± 0.75	0.57 ± 0.78	0.42 ± 0.53
PMN cells	89.2	91.5**	89.6	90.5
Monocytes	10.6	7.9	10.2	9.3
Lymphocytes	0.2	0.6**	0.2	0.2
Six hours				
Total leukocytes				
Viable	31.00 ± 33.23	24.40 ± 18.54	5.20 ± 2.31	8.00 ± 5.47
Dead	4.80 ± 4.40	5.00 ± 5.58	0.60 ± 0.80	1.40 ± 1.96
PMN cells	87.7	85.7	91.8	87.2
Monocytes	12.2	14.1	8.1	12.7
Lymphocytes	0.1	0.2	0.1	0.1

Table 8.2 Pain scores and percentage of degranulated mast cells on pleural surface. The evaluation was done in different times after pleural carrageenan inoculation. (#) Values represented by median and interval. (##) Values represented by percentage. *p = 0.0001, Kruskal-Wallis, in relation to the other groups. X^2, without significance

Time/ parameter	*Arnica montana* Mother tincture	*Arnica montana* 6CH	Dexamethasone	5% hydro-alcoholic solution
One hour				
Pain score (#)	2 (0–3)	1 (0–2)	1 (1–2)*	2 (1–2)
Degranulated mast cells (##)	85.8	83.9	84.3	86.2
Four hours				
Pain score	2 (1–2)	2 (0–2)	1(0–2)	2(1–2)
Degranulated mast cells	87.0	87.5	86.8	89.4
Six hours				
Pain score	1 (1–2)	1 (0–2)	0 (0–2)	2 (1–2)
Degranulated mast cells	92.4	92.3	92.9	91.2

Experiment 2 –Histochemistry and Histometry

In this experiment, the subcutaneous tissues obtained from 3 hours inoculated footpads were processed histochemically, using two special staining techniques: PAS and Verhoeff, in order put in evidence the vessels morphology before measuring their area with an image analysis system (Image Tool 3.0). The calculation of vessels net area average and the establishment of their relation to the edema reduction represented an important parameter to understand the connective tissue behavior during acute inflammation, mainly after *Arnica montana* 6CH treatment. Also, the Toluidine blue method was used again, to challenge the previous data obtained from lung slides. Ten rats were used by group.

The PAS (Periodic Acid Schiff) technique allows mark polysaccharides in tissue structures, as basal lamina. Thus, the outline of vessels became very evident, which makes easier their recognition and measurement by the software (Image Tool 3.0). The Verhoeff technique is based on silver staining of basal lamina and has the same purpose, mainly to identify little arteries in a connective tissue. Three randomized microscopic fields per slide were digitally recorded in this step. All slides were identified by codes; such a manner that image analysis was done in blind. In order to correlate the putative changes in vessels net area with edema reduction, the ratio between average area and paw volume was calculated and expressed graphically.

The percentage of degranulated mast cells was smaller in *Arnica montana* treated animals (mother tincture and 6CH), contrary to the results obtained from lung slides (Table 8.3). However, only animals treated with *Arnica montana* 6CH presented significant edema reduction and higher vessels/edema ratio (Table 8.3, Fig. 8.1). This study was presented in the XXI symposium of GIRI, 2007 (Bonamin, 2007).

Table 8.3 Edema and degranulated subcutaneous mast cells of rats inoculated with 1% carrageenan into the footpad. Time after inoculation = 3 hours. (#) Values represented by mean ± standard deviation, mm^3. (##) Values represented by percentage. $^*p \leq 0.05$, ANOVA/Tuckey Krammer, in relation to hydro-alcoholic solution; $^{**}p \leq 0.02$, X^2, in relation to hydro-alcoholic solution

Time/ parameter	*Arnica montana* Mother tincture	*Arnica montana* 6CH	Dexamethasone	5% hydro-alcoholic solution
Edema (#) (elliptic calculated-volume)	0.085 ± 0.063	0.052 ± 0.050*	0.063 ± 0.046	0.166 ± 0.126
Degranulated mast cells (##)	39.7**	48.7**	53	54

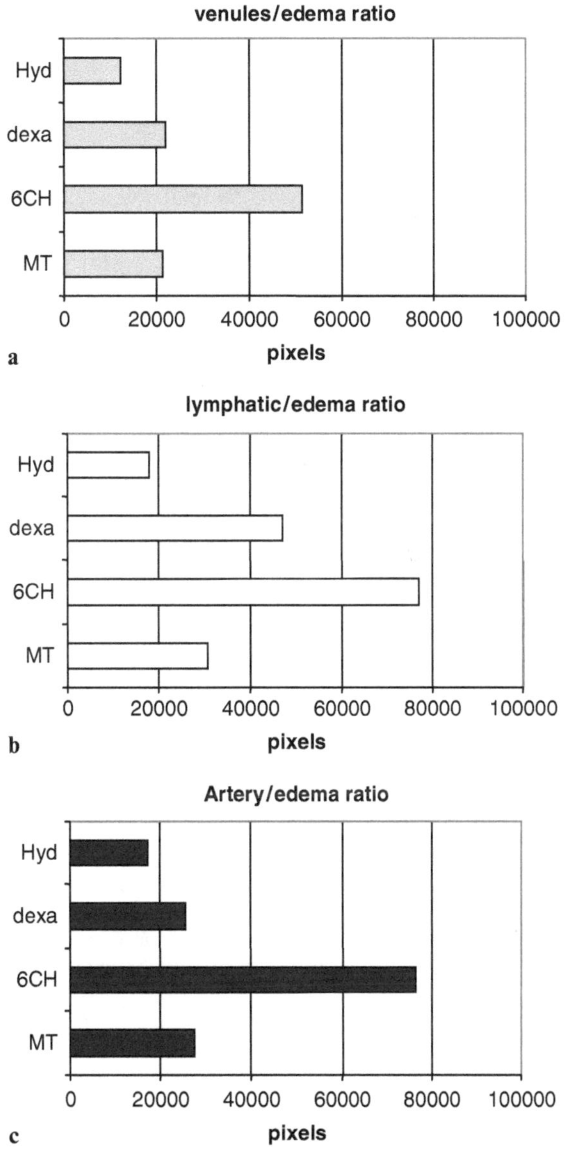

Fig. 8.1 Vessels area/edema ratio in rats inoculated with 1% carrageenan into the footpad. Time after inoculation = 3 hours. Venules, lymphatic vessels and arterioles area average (pixels) were calculated by the software Image Tool 3.0. Hyd = 5% hydro-alcoholic solution; dexa = dexamethasone; 6CH = *Arnica montana* 6CH; MT = 5% mother tincture *Arnica montana*

Experiment 3 – Lymphatic Absorption

In this step, a specific model was used to verify if *Arnica montana* 6CH treated animals presented changes in lymphatic absorption of inflammatory edema. Details about the method are described above. Thirteen rats were used by group. Preliminary results of this experiment are seen in Ferrari and Bonamin (2006).

Animals treated with *Arnica montana* 6CH presented significant increased lymphatic absorption in relation to control. Dexamethasone treated animals presented mild increase in this parameter (Table 8.4).

Experiment 4 – Comparison Between Two Different Potencies

Considering the observed modulator effects of *Arnica montana* 6CH upon the acute inflammatory process, a decimal potency of *Arnica montana* was also tested, using the same experimental model. Ten rats were used by group. The aim of this step was to compare the 6CH potency – which is very used to acute situations – with the 3D potency, which is also very used in clinical practice, and verify if the same effects were present in this case. Edema and subcutaneous degranulated mast cells counting after 3 hours from carrageenan inoculation were analyzed in this experiment. The *Arnica montana* D3 and the vehicle were obtained from *Weleda do Brasil* Laboratory. The mast cell counting, edema and statistical analysis were performed as described above. Details about this study were related in Bittencourt et al. (2002).

Table 8.4 Local lymphatic absorption of edema from rats inoculated with 1% carrageenan into the footpad. Time after inoculation = 3 hours. Values represented by mean ± standard deviation, μg/ml. *p ≤ 0.05, ANOVA/Dunnet, in relation to hydro-alcoholic solution

Time/ parameter	*Arnica montana* Mother tincture	*Arnica montana* 6CH	Dexamethasone	5% hydro-alcoholic solution
Evans blue stain concentration in formamide (μg/ml)	181.71 ± 78.44	293.95 ± 171.83*	211.20 ± 36.03	179.19 ± 86.89

Table 8.5 Edema and degranulated subcutaneous mast cells of rats inoculated with 1% carrageenan into the footpad. Time after inoculation = 3 hours. (#) Values represented by mean ± standard deviation, mm^3. (##) Values represented by percentage. *p ≤ 0.05, ANOVA/Tuckey Krammer, in relation to vehicle; **p = 0.0001 X^2, in relation to vehicle

Time/ Parameter	Arnica montana 3D	Dexamethasone	Weleda vehicle
Edema (#) (elliptic calculated volume)	0.043 ± 0.034*	0.058 ± 0.037*	0.101 ± 0.024
Degranulated mast cells (##)	45.2**	58.2**	66.9

The results showed that *Arnica montana* D3 also reduced the edema formation in a magnitude comparable to dexamethasone and the percentage of degranulated mast cells was reduced in the same proportion (Table 8.5).

Discussion

One of the difficulties to understand the mechanisms related to Arnica treatment is the lack of available literature. In classical studies, Campbell (1976) showed in human volunteers, the effects of Arnica in induced edema. Michaud (1987) described that *Arnica montana* 15CH, together with *Apis mellifica* 7CH, was able to reduce post-surgical edema. Recently, Oberbaum et al. (2005) observed that treatment with homeopathic *Arnica montana* is effective to reduce post-partum blood loss; Brinkhaus et al. (2006) showed reduction of edema secondary to knee surgery after treatment with *Arnica montana* 30x and Robertson et al. (2007) related significant reduction of pain in post-tonsillectomy with *Arnica montana* 30C treatment. Moreover, the use of Arnica has been considered as an interesting possibility in plastic surgery: Totonchi and Guyuron (2007) demonstrated that Arnica and steroids are both effective to reduce postrhinoplasty edema and Seeley et al. (2006) showed significant reduction in edema and ecchymosis after face-lifts.

In fact, most part of scientific articles that have already been published is related to clinical observations, with no focus on physiopathology.

Herein, the modulation effects of *Arnica montana* 6CH upon acute inflammation were put in evidence. Preliminary results revealed that reduction of carrageenan induced inflammatory edema by Arnica is related to histamine and prostaglandin participation, even without changes in vascular permeability. These findings motivated the research of putative modulation effects of Arnica, instead of classical anti-inflammatory mechanisms.

Carrageenan subcutaneous or intra-pleural inoculations are classical experimental models to induce acute inflammation. Thus, its patterns of edema, vascular permeability and cell migration are well known. For this reason, this model provides consistent references to study the effects of substances that interfere with inflammatory process, mainly with chemical mediators. Histamine, serotonin, prostaglandin and bradikinin are important chemical mediators involved in carrageenan induced acute inflammation (Rocha and Garcia, 1972). Recently, this method was used as a controlled way to verify the effects of homeopathic remedies on inflammation and to discuss some methodological difficulties that are often present in this kind of study (Conforti et al., 2007).

The results obtained after pleurisy induction point toward the persistence of PMN cells migration into the inflammatory site (even after 4 hours after inoculation) together with early lymphocyte migration, reflecting the array of the acute phase and the trend to improve specific cellular response against the etiological agent in animals treated with *Arnica montana* 6CH. This framework is quite different from that shown by dexamethasone, in which decrease in leukocyte migration and increase of *in situ* leukocyte degeneration were very clear. Glucocorticoids have a wide range anti-inflammatory mechanism, which includes inhibition of several chemical mediators, leukocytes inhibition and degeneration (Bonamin and Paulino, 1999). Thus, the results strongly suggest that dexamethasone and *Arnica montana* 6CH play different games, acting in different manners, although reduction of pain behavior was observed in both cases. As pain behavior can be related to several physiological conditions, it is possible that these changes could be an indirect consequence of exudative and cellular modulation in rats treated with *Arnica montana* 6CH.

The percentage of degranulated mast cells presented different outputs according to the inflammation site. In the pleurisy model, no change in mast cell degranutation was seen. Instead, in paw subcutaneous model, a clear reduction of mast cell degranulation could be observed, together with edema reduction. This framework was also seen after *Arnica montana* D3 treatment. The reasons to explain these discrepancies are unknown but invite further studies about leukocyte dynamic in subcutaneous inflammation model. Anyway, changes in membrane stability were already described for Arnica extracts (Teske and Trentini, 1994) suggesting that this can be another parameter to be related to Arnica treatment or proving.

In order to know better the vascular and connective tissue behavior of subcutaneous tissue in animals treated with *Arnica montana* 6CH, two other models were applied. The histochemical study revealed increase in vascular area, mainly arterioles and lymphatic vessels, related to *Arnica montana* 6CH. The relation between vessel area and edema reduction was very clear in these animals. This finding support the hypothesis that edema absorption is one important mechanism involved in edema reduction, more than changing in vascular permeability, as described above.

The data are compatible with the initial hypothesis that the mother tincture of *Arnica montana* would act locally by its irritant properties, mainly in dermal tissues (Carvalho, 2000; Macedo et al., 2004; Jancovic et al., 2004), in such a manner that it would stimulate the local blood flow and, consequently, could establish local symptoms that would have tight similitude with those induced by carrageenan (Rocha and Garcia, 1972; Conforti et al., 2007). The active principles responsible for this action are lactones-sesquiterpenes, in which helenalin is included (Wordenbarg et al., 1994; Hausen, 1996; Spetolli et al., 1998; Kos et al., 2005). Bradikinin, prostaglandins, serotonin and cytokines are tissue mediators involved in helenalin biological activity (Schröder et al., 1990; Powis et al., 1994; Klaas et al., 2002; Jäggi et al., 2004).

In this sense, a last experiment was done, in order to provide information about lymphatic absorption dynamic in treated rats, whose result confirm the hypothesis of higher absorption.

Conclusion

The data, taken together, highlight scenery of more dynamic connective tissue in response injuries, with more active PMN cell migration and more intense blood flow and lymphatic edema absorption. This scenery is compatible with the *similia principle* but inconsistent with classical anti-inflammatory activities, as seen in dexamethasone treated rats.

Acknowledgements We thanks to the *Vice-Reitoria de Pós-Graduação*, University Paulista and FAPESP (Proc. 97/4810-2; 99/09274-7; 04/15631-0) for support; to *Weleda do Brasil* that gently provide us all the necessary material and grants for some students; to Paulo Ailton Valdovato for making and staining slides and, mainly, to all the students and colleagues involved in this study, namely Aloísio Cunha de Carvalho, Mariana Lage Marques, Letícia Cardoso Bittencourt, André Dias Diegues and Fernanda Samapaio Ferrari.

References

Bittencourt LC, Dieguez AD, Carvalho AC. Efeitos da administração de *Arnica montana* (tintura-mãe e preparações UHD) nas alterações estruturais vasculares durante o processo inflamatório agudo. VII SINAPIH. May 2002; Rio de Janeiro, Brazil.

Bonamin LV. *Arnica montana* and behavior of connective tissue: 10 years of study. XXI Symposium of GIRI. March, 2007; Monte Carlo, Monaco.

Bonamin LV, Paulino CA. Imunofarmacologia. In: Spinosa HS, Górniak SL, Bernardi MM (Eds.). Farmacologia aplicada a Medicina Veterinária. 2ed. Rio de Janeiro: Guanabara Koogan; 1999.

Bonamin LV, Lage-Marques M, Carvalho AC. Efeitos da *Arnica montana* (tintura-mãe e ultradiluições) no processo inflamatório agudo. Reunião anual da FESBE. August, 2000; Caxambu, Brazil.

Brinkhaus B, Wilkens JM, Ludtke R, Hunger J, Witt CM, Willich SN. Homeopathic arnica therapy in patients receiving knee surgery: results of three randomized double-blind trials. Complement. Ther. Med.; 2006, 14: 237–246.

Bucay JW. Algumas notas sobre la planta medicinal *Arnica montana*. Revista Médica del Instituto Mexicano del Seguro Social; 1955, 33: 306–326.

Campbell A. Two pilot controlled trials of *Arnica montana*. Br. Hom. J.; 1976, 3: 154–158.

Carvalho AC. Efeitos da administração de Arnica montana (tintura-mãe e UHD) na atividade de diferentes agentes flogísticos em ratos, Brazil: Dissertação de mestrado: Universidade Paulista; 2000.

Carvalho AC, Bonamin LV. Efeitos da *Arnica montana* (tintura-mãe e preparações dinamizadas 6CH) na atividade de diferentes agentes flogísticos em ratos. Pesquisa Homeopática; 2001, 16: 31–43.

Cerqueira M, Souza J, Junior R, Peixoto A. A ação analgésica do extrato bruto aquoso liofilizado do caule e folhas da *Lychnopora ericoides* (arnica). Ciência e Cultura; 1987, 39: 551–553.

Chakrabarti A. Treatment of haematoma in dogs: a non surgical approach. Indian Vet. J.; 1991, 71: 1229–1230.

Chakrabarti J, Biswas SJ, Khuda-Bukhsh AR. Cytogenetical effects of sonication in mice and their modulations by actinomycin D and homeopathic drug, Arnica 30. Indian J. Exp. Biol.; 2001, 39: 1235–1242.

Conforti A, Bellavite P, Bertani S, Chiarotti F, Menniti-Ippolito F, Raschetti R. Rat models of acute inflammation: a randomized controlled study on the effects of homeopathic remedies. BMC Complement. Altern. Med.; 2007, 7: 1. DOI: 10.1186/1472-6882-7-1.

Cotran RS, Kumar V, Collins T. Pathologic Basis of Disease – Robbins, 7th edition. Philadelphia, PA: Saunders Company WB; 2004.

Demarque D, Jouanny J, Poitevin B, Saint-Jean Y. Homeophathie, connaitre la materie medicale. CEDH; 1985, 1: 17–21.

Ernst E, Pittler MH. Efficacy of homeopathic Arnica – A systematic review of placebo-controlled clinical trials. Arch. Surg.; 1998, 133: 1187–1190.

Ferrari FS, Bonamin LV. Efeitos da *Arnica montana* (6CH, 12CH e Tintura-mãe) na Reabsorção Linfática de Ratos Submetidos a Modelo de Inflamação Aguda. XX Symposium of GIRI. September, 2006; São Paulo, Brazil.
Genet L. Arnica – Grandes characteristiques: contusion, douleur, hipersensibilitée. Les annals Homeophathiques Française; 1980, 4: 91.
Giraldi T, Perissin L, Zorzet S, Rapozzi V. Stress, melatonin and tumor progression in mice. Ann. N.Y. Acad. Sci. USA; 1994, 719: 526–535.
Hall IH, Lee KH, Starnes CO, Muraoka O, Sermida Y, Waddel TG. Antihyperlipidemic activity of sesquiterpene lactones and related Compounds. J. Pharm. Sci.; 1980, 69: 694–696.
Hausen BM. A 6-year experience with compositeae mix. Am. J. Contact Dermatitis; 1996, 7: 94–99.
Jäggi R, Würgler U, Grandjean F, Weiser M. Dual inhibition of 5-lipoxygenase/cyclooxygenase by a reconstituted homeopathic remedy; possible explanation for clinical efficacy and favourable gastrointestinal tolerability. Inflamm. Res.; 2004, 53: 150–157.
Jancovic M, Poljacki M, Duran V, Vuanovic L, Sente R, Stojanovic S. Contact allergy to Compositae plants in patients with atopic dermatitis. Med. Pregl.; 2004, 57: 209–218.
Jeffrey SL, Belcher HJ. Use of Arnica to relieve pain after carpal-tunnel release surgery. Altern. Ther. Health Med.; 2002, 8: 66–68.
Klass CA, Wagner G, Laufer S, Sosa S, Della Loggia R, Bomme U, Pahl HL, Merfort I. Studies on the anti-inflammatory activity of phytopharmaceuticals prepared from Arnica flowers. Planta Med.; 2002, 68: 385–391.
Kos O, Lindenmeyer MT, Tubaro A, Sosa S, Merfort I. New sesquirterpene lactones from Arnica tincture prepared from fresh flowerheads of *Arnica montana*. Planta Med.; 2005, 71: 1044–1052.
Lyss G, Schimidt TJ, Merfort I, Pahl HL. Helenalin, as anti-inflammatory sesquiterpene lactone from Arnica, selectively inhibits transcription factor NF-κB. Biol. Chem; 1997, 378: 951–961.
Macedo SB, Ferreira LR, Perazzo FF, Tavares Carvalho JC. Anti-inflammatory activity of *Arnica montana* 6CH: preclinical study in animals. Homeopathy; 2004, 93: 84–87.
Michaud J. Action d'Apis mell et d'Arnica dans la prevention des oedemes pos-opératoires en chirurgie maxilo – faciale à propos d'une expérimentation clinique sur 60 observations. In: Poitevin B (Ed.). Le devenir de l'Homeopathie – Elements de theorie et recherche. Paris: Doin Editeurs; 1987.
Moriguchi P, Sannomiya P, Lara PF, Oliveira-Filho RM, Greco KV, Sudo-Hayashi LS. Lymphatic system changes in diabetes mellitus: role of insuline hyperglycemia. Diabetes/Metab Res. Rev.; 2004, 21: 150–157.
Oberbaum M, Schreiber R, Rosenthal C, Itzchaki M. Homeopathic treatment in emergency medicine: a case series. Homeopathy; 2003, 92: 44–47.
Oberbaum M, Galoyan N, Lerner-Geva L, Singer SR, Grisaru S, Shashar D, Samueloff A. The effect of the homeopathic remedies *Arnica Montana* and Bellis perennis on mild postpartum bleeding – A randomized, double-blind, placebo-controlled study-Preliminary results. Complement. Ther. Med.; 2005, 13: 87–90.
Powis G, Gallegos A, Abraham RT, Ashendel CL, Zalkow LH, Grindey GB, Bonjouklian R. Increased intracellular Ca2 + signaling caused by the antitumor agent helenalin and its analogues. Cancer Chemother. Pharm.; 1994, 34: 344–350.
Ramelet AA, Buchheim G, Lorenz P, Imfeld M. Homeopathy Arnica in postoperative haematomas: a double-blind study. Dermatology; 2000, 201: 347–348.
Rigamonti B. Arnica a magical remedy against any type of trauma. Obiettivi e Documenti Veterinari; 1995, 16: 67–68.
Robertson A, Suryanarayanan R, Banerjee A. Homeopathic *Arnica montana* for post-tonsillectomy analgesia: a randomised placebo control trial. Homeopathy; 2007, 96: 17–21.
Robles M, Aregullin M, West J, Rodriguez E. Recent studies on the zoopharmacognosy, pharmacology and neurotoxicology of sesquiterpene lactones. Planta Med.; 1995, 61: 199–203.
Rocha S, Garcia L. Chemical mediators of the acute inflamatory reactions. New York: Pergamon; 1972.

Schimidt T, Willuhn G, Wendisch D, Steigel A. A 7(12) 6-trans-lactonzed helemanolide lignas and inositol esters from *Arnica alpina* sp. attenuata. Planta Med.; 1993, 59: 598–599.

Schröder H, Lösche W, Strobach H, Leven W, Willuhn G, Till U, Schrör K. Helenalin and alpha,13-dihydrohelenalin, two constituents from *Arnica montana* L., inhibit human platelet function via thiol-dependent pathways. Thromb. Res.; 1990, 57: 839–845.

Seeley BM, Denton AB, Ahn MS, Maas CS. Effect of homeopathyc *Arnica montana* on bruising in face-lifts: results of a randomized, double-blind, placebo-controlled clinical trial. Arch. Facial Plast. Surg.; 2006, 8: 54–59.

Spetolli E, Silvani S, Lucente P, Guerra L, Vincenzi C. Contact dermatitis caused by sesquiterpene lactones. Am. J. Contact Dermatitis; 1998, 9: 49–50.

Stevinson C, Devaraj VS, Fountain-Barber A, Hawkins S, Ernst E. Homeopathic Arnica for prevention of pain and bruising: randomized placebo-controlled trial in hand surgery. J. R. Soc. Med.; 2003, 96: 60–65.

Teske M, Trentini MM. Herbarium-Compendio de Fitoterapia, Curitiba: Edição Herbarium Laboratório Botânico; 1994.

Totonchi A, Guyuron B. A randomized, controlled comparison between arnica and steroids in management of postrhinoplasty ecchymosis and edema. Plast. Reconstr. Surg.; 2007, 120: 271–4.

Wolf M, Tamaschke C, Mayer W, Heger M. Efficacy of Arnica in varicose vein surgery: results of a randomized, double-blind, placebo-controlled pilot study. Forsch. Komplementarmed. Klass. Naturheilkd.; 2003, 10: 242–247.

Wordenbarg HJ, Merfort I, Passruter CM, Van U. Cytotoxicity of flavonoides and sesquiterpene lactonas from Arnica species against the GLC4 and COLO 320 cell lines. Planta Med.; 1994, 60: 434–437.

Chapter 9
How Living Organisms Detect Very Low Doses? A New Approach to Homeopathy and Hormesis

Elżbieta Malarczyk

Processes Regulating Cellular Metabolism

The molecular strategy of integrating cellular processes is based on the controlling mechanisms of the two major opposing pathways: catabolism and anabolism. Owing to this, there is a constant exchange of substrates and products of metabolism, which provide cells with indispensable elements for their survival in specific environmental conditions. The control of speed and direction of transformations in a cell is based on protein systems. They combine into complexes, such as receptors and multi-enzymatic complexes, thanks to which the signal gets transmitted both from the gene to the protein and then to the environment and the other way, having a precise effect on the regulation and efficiency of all cellular processes (Welch, 1995; Welch and Clegg, 1987). The transmission proceeds gradually by electron exchange at the level of chemical bonds in a series of cyclic or cascade reactions catalyzed by specialized protein systems, which, as a consequence of receiving information in this way, change their spatial conformation (Ho, 2005).

One of the most frequently used mechanisms of information reception is structural allosteric adjustment within one or several protein subunits. That is why the multi-stage activity regulation of regulatory proteins works as a biological trigger, which is particularly sensitive to subtle changes in the concentration of specific low-molecular effectors, present in the environment (Malarczyk and Pazdzioch-Czochra, 2000). The basis of the triggering mechanism is the ability of the allosteric protein to occur in two separate conformational states referred to as an open or a closed state depending on the presence of inhibitors or activators. Thanks to this unique property, changes in the direction of a certain reaction can occur almost instantly, even after a minor change in the concentration of effectors (Andersen and Doving, 1991).

All cell regulatory processes are strictly dependent on its energy level. Every living organism possesses molecular power plants responsible for the production

M. Curie-Sklodowska University, Lublin, Poland. malar@poczta.umcs.lublin.pl

Presented part at the XIXth Symposium of GIRI, Monaco, 2005 and part at the XXIth Symposium of GIRI, Monaco, 2007

and distribution of energy (Liburdy, 1994). In animal organisms, energy is produced in mitochondria, whose membranes are equipped with elements of the respiratory chain, owing to which protons and electrons originating in central metabolism are constantly transported onto atomic oxygen. This is accompanied by synthesis of water molecules (Luzar and Chandler, 1996), without which life would be virtually impossible, and by production of high energy ATP, indispensable for all cellular syntheses. ATPase, which is responsible for the production of ATP, works like proton pomp and is an example of a highly sophisticated molecular mechanism which transforms a stream of electrons into high energy bonds stored, precisely, in ATP. The direction of electrons and protons flow also determines the speed of formation or breakdown of this compound, which allows further ion exchange and, ultimately, sends messages over large distances through neural networks, i.e. in cellular respiratory systems, efficient electronic flow is also taken care by numerous metalloproteins, among which enzymes from the class of oxidoreductases are particularly numerous. They facilitate the ultimate combination of hydrogen and oxygen thanks to electron exchange. They are also responsible for removing the side metabolites of this same combination, such as hydrogen peroxide or reactive forms of oxygen called free radicals. Those highly reactive molecules endowed with an additional electron significantly affect the regulation of respiratory processes in cells, especially if they occur in excessive numbers, thus, showing a harmful oxygen effect. Co-operation between hydrogen and oxygen also manifests itself in the formation of numerous hydrogen bonds responsible for the appropriate spatial form of biopolymers.

Water is an irreplaceable cell medium, it means: the product of joining together hydrogen and oxygen. It appears in cells permanently, not only as an outcome of respiratory processes, but it also accompanies many other reactions, both synthetic and hydrolytic. It also serves as a proper medium for processes related with cell polarity and electron transfer. However, its putative ability to remember information about the chemical structure of a substance dissolved in it is still a matter of contention between supporters of homeopathy and its opponents.

Reactions accompanying biological oxygen protonation are still not fully known. There is also no explanation of how oxydoreductases reduce the high exothermic oxidation of organic molecules and how oxygen is activated in them (Rey, 2003). It is common knowledge that the process of combustion is characterized by an abrupt energy release and a high ignition temperature. During digestion of food, living organisms can channel exothermic processes and store surplus energy from combustion in ATP molecules. Although these transformations are equivalent to combustion, the oxygenase enzymes which control specific reaction pathways release energy gradually. We owe the formation of oxygen metabolism to evolutionary processes.

Thanks to this mechanism, gradual combustion of food provides the organism with a maximum amount of energy stored as glucose, carbohydrates and fats. When there is sufficient supply of oxygen, glucose transforms into derivatives of pyruvic acid and the tricarboxylic acid cycle (Krebs) of the respiratory chain gets combusted into water and carbon dioxide. However, when oxygen is lacking, cells switch to anaerobic metabolism, the main product of which is lactic acid. The latter does not enter the transformations of the Krebs cycle, the glucose is used for low energy transformations with the omission of the respiratory chain. The most recent research shows that the decoupling of *glycolysis-Krebs cycle-respiratory chain system* lies at the basis of very expansive growth of tumors, and it is the poorly working, "lazy" mitochondria that are blamed for this. It has been proven that these organelles are inefficient in cancer tissues and give a much weaker coupling between glycolysis and the respiratory chain. Such transformation is less energy-effective and additionally produces harmful metabolites, which make it impossible for a cell to enter into apoptosis. It is estimated that in the near future the use of lactate fermentation inhibitors may replace chemotherapy in the battle against cancers (http://www.depmed.ualberta.ca/dca/letter_031507.htm).

This brief review about the processes responsible for the regulation of cell metabolism at the molecular level leads to the conclusion that all the processes are based on electron relations between the hydrogen and oxygen atoms. That is why the search for an answer to how strongly diluted substances influence living systems should be connected to the observation of enzymatic respiratory cell systems behavior.

The Regulatory Role of Oxygenases

Similarly to the enzymes of the respiratory chain, other cell oxygenases actively join in electron transformations in accordance with the structure of their catalytic centers, rich mainly in ferric or cupric ions. Commonly occurring oxidases, oxygenases or peroxidases protect strategically important cell elements against the consequences of rapid overly of potentials difference, using for this purpose the gradual transfer of electrons in the catalytic center and in the accompanying protein ligands. Among the many protein systems, the enzyme peroxidase is a very effective research object. That is because it has all attributes of a regulatory enzyme, and simultaneously possesses a heme center, exhibiting the characteristics of a biological oscillator (Bronnikova et al., 2001; Schaffer et al., 2001). For this reason, three enzymes from the group of oxydoreductases which accompany aerobic transformations of phenolic substances – aromatic compounds rich in hydroxyl or methoxyl groups – were chosen as a model for the studies about the response of protein systems to the presence of highly diluted effectors at the molecular level (Olsen et al., 2003). Phenolic substances have high affinity for fungi and some actinomycetes, since they constitute products of decomposition of the natural biopolymer lignin. The selected enzymes were laccase, fungal and plant peroxidase, and O-demethylases from actinomycetes of the Nocardia group. Spectroscopy and luminescence methods, electrophoretic tests and electron microscopy were mainly used for monitoring morphological changes in the examined cells. Moreover, working with purified plant peroxidase protein, it is possible to track comfortably the dynamics of transformations induced by the presence of highly diluted phenolic effectors in the reaction environment.

Observations of Information Reception Mechanisms by Organisms Containing Highly Diluted Effectors

Experimental evidence concerning the reaction of fungi to the presence of diluted effectors has already been published (Malarczyk et al., 2003). Additionally, studies of pure HR-peroxidase (Malarczyk et al., 2004) and O-demethylases from *Nocardia sp* have enriched our knowledge about the response of enzymatic protein according to changes in the concentration of low-molecular co-reagents. Overall conclusions could be presented, which point to the principles obtaining from the biochemistry of very low doses. Because the dilutions of phenolic substances were prepared in accordance to the principles of homeopathy, the studies may throw new light on the controversial aspects of the action of homeopathic drugs.

Although this subject is very new and needs more and more investigation, some commentaries and conclusions that could be drawn from a preliminary analysis of the material collected by the present author and other authors are presented:

1. *During potentization (shaking) of dilutions, there might be a gradual inverse relation between matter and energy. This shift might be critical to define the steady information about the nature of the diluted substance, even after successive dilutions.*

The graphic illustration of the progressing dilution resembles the model of the Fibonacci spiral. Evidence for disturbing energy in the diluted solutions has been provided in a number of experimental studies by Elia and coworkers (Elia et al., 2006; Elia and Niccoli, 1999, 2000) who have shown, using the calorimetric and the conductometric methods, higher level of heat and changes in conductivity in potentized solutions of salts, acids, alkali, and a number of organic compounds.

The increase in the level of kinetic energy in a solution manifests itself by the emission of light. That is why the use of luminometry and spectrophotometry to detect the effects of individual dilutions gives the possibility to show differences in the energetic pattern of different dilutions (Malarczyk, 2007a; Malarczyk et al., 2007b).

2. *Preparation of successive dilutions is connected with selection of an iteration value (1:10, 1:100, 1:1,000, etc.) and requires constant repetition of the potentization procedure, from the starting to the moment when the desired potency has been obtained. It is assumed that information concerning the action of the initial substance would become encoded in the individual solutions, thus each individual dilution could be used to continue the process of potentization.*

Potentized or dynamized dilutions can be prepared from any chemical substance or from herbal extracts, tissue preparations, etc., using the technique of successive transfers. Preparation of a homeopathic drug is, on principle, based on making series of dilutions of different biologically active substances (Dr. Schwabe "Pharmacopoeia Homeopathica Polyglotta", 1872) followed by potentization by rhythm shaking. Diluting often significantly exceeds the Avogadro constant; however, the individual dilutions seem to preserve the information about the matrix-substance, showing characteristic physiological properties, often opposite to the effect of the initial dose.

3. *Potentized solutions might have effects on selected biological systems-receptors, multi-enzymatic complexes, specific allosteric enzymes, etc., in accordance with the principle of affinity between the effectors and complex.*

The literature about homeopathic dilutions shows that the recognition of individual solutions by physical-chemical methods has practically not been found. However, in biological studies, one can clearly observe different effects of active substances in dependence on the dilution level. Among numerous examples, worth citing are studies on the two-phase response of specific receptors by changing amounts of dexamethasone (Bonamin et al., 2001) or different reactions to high and low doses of glutamate (Jonas et al., 2001). In studies of tadpole metamorphosis, Endler proved strong connection of the process with selected dilutions of thyroxin (Endler and Schulte, 1994; Endler et al., 1995). Inhibition of basophile degranulation by histamine dilutions as well as the antitrombotic activity of aspirin dilutions were proofed and wide commented by researchers from France (Belon, 2006). Also, homeopathic therapy is known for its impact on a specific set of organic characteristics. It is likewise according to the specificity of their action that isopathic drugs are also selected; the aim of which is to remove toxin excess from organisms (animal or plants) by means of its higher dilutions (Betti et al., 1997, 2003).

Similar evidence has been provided by studies about the relation between oxidoreductases and their cofactors in fungi from *Basidiomycetes* class (Malarczyk et al., 2003). Support has been found in experiments with peroxidase and various phenolic substances (Malarczyk et al., 2004), laccase and its substrates, and O-dioxygenases in *Nocardia* (Actinomycetes), in response to different dilutions of formaldehyde (Malarczyk and Pazdzioch-Czochra, 2000). All those substances showed affinity for a specific type of reaction in enzymatic catalysis.

4. *Individual serial dilutions might differ in their action on biological systems.*

A great symmetry of the rhythmical time-changeable responses of cells and isolated enzymatic proteins to the presence of gradually diluted effectors has been observed in the studies cited above. It points to a periodic character of the observed transformations, where the extreme states of the studied systems determine their maximum and minimum activity, as seen in wave-shaped graphs (Malarczyk, 2007a, 2008; Malarczyk et al., 2007b).

In fungi, potentized phenols have been observed to produce clear differences in enzymatic activity, as well as increase and decrease in protein production; for pure plant peroxidase from horseradish, transition from oxidase activity to peroxidase activity has been observed. In *Nocardia*, disappearance and appearance of vacuolization in cells as well as periodic abrupt secretion of oxygen depended on the level of dilution of formaldehyde. Multiple, repeated results constantly showed the same values of respective dilutions responsible for extreme function activation or inhibition of the biological detector (Malarczyk, 2007a; Malarczyk et al., 2007b).

5. *Since the type of shaking might influence some parameters, it is advisable to always use the same potentization procedure, uniform for all dilutions.*

Mechanic dynamization, with a laboratory shaker, for instance, is also appropriate if the right speed is used (Elia and Niccoli, 2000). Dynamization is the basis for preparing a homeopathic drug.

6. *Oscillation between extreme states of activation or inhibition of biological activity might depend on the degree of dilution.*

For HR-peroxidase, i.e., the maximal points on sinusoidal curve correspond to high peroxidase activity and minimal to high oxidase activity (Malarczyk, 2007, 2008; Malarczyk et al., 2007). It could be emphasized that the inhibition of basophile degranulation by histamine (Belon, 2006) reveal the same number of dilutions responsible for maximal or minimal % of degranulation as it was observed in experiments with fungal material and HR-peroxidase.

7. *A wave graph could be made up from individual dilutions of a given substance, which preserve the information about its type in a permanent way, repeatable with reference to the studied system.*

The periodic character of the changes evoked by homeopathic doses has been pointed out for years in works of numerous researchers, among them Delinick (1991), Bastide (Bastide et al., 1987; Bastide and Boudard, 1995; Bastide, 2002), Betti (Betti et al., 1997, 2003), Endler (Endler and Schulte, 1994; Endler et al., 1995), Bonamin (Bonamin et al., 2001) and many others, who have become pioneers of the scientific bases of homeopathy. The current view of experimental results on the effect of high dilutions in diverse biological material is a continuation of the predecessors thought, despite it introduces new elements by pointing out the essential role of biological systems capable of existing in opposing conformational states, according to high dilutions exposition.

The fact that wave graph have the length measured in consonance with the degree of dilution is also a new knowledge and deserves further studies. Currently, there is the possibility of experimentally determining the length of a wave for any substance and testing its relation with a compatible receptor system for all living organisms.

8. *The graph waves could change its pattern under the influence of perturbations, such as extreme temperature.*

The experimental proofs brought much information about the wave lengths of diluted phenolics, formaldehyde, glycine, betaine and others N-methylated active compounds during their incubation with fungal, bacteria and plants, respectively (Bellavite et al., 1999; Malarczyk et al., 2003, 2004; Schulte and Endler, 1998). For animal and human material, systematic assays are still lacking.

9. *Enzymes from the group of oxidoreductases change according to electronic transformations of oxygen and protons.*

As has been shown in studies using phenols or their acids, where the presence of free hydroxyl or carboxyl groups is clearly marked, the dynamics of oscillation remains undisturbed, even when the dilution exceeds 100^{-30} (Malarczyk et al., 2004).

It is enough, however, to change the phenol OH group to the less reactive OCH_3 group in anisol or veratric acid to cause a gradual suppression of oscillation (Hauser and Olsen, 1998). This, at least partly, could explain the *mystery* of water memory. It still amazes people and points toward the possibilities of dynamic adaptation of water to the environmental conditions, based on O-H bond resonance. These very important statements lead to practical strategy in anticancer prevention (Bell et al., 2003; Biswas and Khuda-Bukhsh, 2002; Brown et al., 2003; Motwani et al., 1992).

10. *The possibility of establishment of selective concentrations and dilutions of biologically active substances with reference to the goal of their action might be considered.*

These dependencies are well-illustrated by the works of Tyihak et al. (Tyihak et al., 2006) on obtaining plant immunity to pathogenic infections by using diluted biological protective agents (Tyihak, 2007). Systematic studies showed sinusoidal changes of plant immunity.

11. *Putative recognition of the periodic character of dilutions could help to account for the concept of overcompensation known from studies of hormesis.*

The periodical character of the studied transformations could be maintained during systematic dilution, even beyond the value of the Avogadro constant. This novel approach might constitute an experimental bridge between the effect of hormesis, isopathy and homeopathy. Many phenomena accompanying the homeopathic procedure are difficult to interpret. These include, among others, the duality of the action of biologically active substances and often seemingly paradoxical effects evoked by their dilutions.

The foundations for understanding the mechanism of hormetic overcompensation are found in the works of Calabrese and Baldwin (Calabrese and Baldwin, 2001; Calabrese, 2001). This relationship has also been used as an argument in prognosis about the effects of radiation catastrophes and other cataclysms connected with environmental contamination: the substance which is poisonous in high doses, after appropriate dynamic dilution might show a totally opposite effect and, as has been experimentally demonstrated, could remove excess poison from the organism (Jonas et al., 2001).

Why it is possible to successfully use both the pharmacological form of a medicine and its selected dilutions? Knowing the length of the corresponding wave and its oscillatory course, one could predict which dilution in a series would reproduce the therapeutic effect of the initial dose and which would inhibit it (Betti et al., 1997, 2003). The knowledge of sinusoidal wave shape points of a certain substance could help to avoid the use of some dilutions in practice. Non-observance of these rules might lead to deterioration of the health condition, upsetting the overall homeostasis or lowering the yield in plant production (Tyihak et al., 2006; Tyihak, 2007).

The evidence of molecular transformations evoked in biological systems of microorganisms and plants with the use of homeopathic or isopathic dilutions gives realistic grounds for continuing the research with reference to people and animals. Systematic reading of characteristic lengths of wave, measured after dilution for numerous medicines, including allopathic drugs, might provide foundations for developing modern, environmentally-friendly nano-pharmacology.

References

Andersen O, Doving KB. Gonadotrophin releasing hormone, a novel olfactory stimulant in fish. Neuroreport; 1991, 2: 458–460.

Bastide M. Recherche Fondamentale. In: Sarembaud A (Ed). Abrégé d'Homéopathie. Paris: Masson Paris; 2002.

Bastide M, Boudard F. A novel concept of immunomodulation. In: Guenounou M (Ed). Forum sur l'Immunomodulation. Paris: John Libbey; 1995.

Bastide M, Daurat V, Doucet-Jaboeuf M, Pelegrin A, Dorfman P. Immunomodulatory activity of very low doses of thymulin in mice. Int J Immunother; 1987, 3: 191–200.

Bell IR, Lewis DA, Brooks AJ, Lewis SE, Schwartz GE. Gas discharge visualisation evaluation of ultramolecular doses of homeopathic medicines under blinded, controlled conditions. J Altern Complement Med; 2003, 9: 25–38.

Bellavite P, Andrioli G, Lussignoli S, Bertani S, Conforti A. Homeopathy in the perspective of scientific research. Ann Ist Super Sanita; 1999, 35: 517–527.

Belon P. (Ed.). Recherche en homeopathie - 4th ed. Lyon, Editions Boiron; 2006.

Betti L, Brizzi M, Nani D, Peruzzi M. Effect of high dilutions of *Arsenicum album* on wheat seedlings from seeds poisoned with the same substance. Br Hom J; 1997, 86: 86–89.

Betti L, Lazzarato L, Trebbi G, Brizzi M, Calzoni GL, Borghini F, Nani D. Effects of homeopathic arsenic on tobacco plant resistance to tobacco mosaic virus. Theoretical suggestions about system variability, based on a large experimental data set. Homeopathy; 2003, 92: 195–202.

Biswas S, Khuda-Bukhsh A. Effect of homeopathic drug, *Chelidonium*, in amelioration of p-DAB induced hepatocarcinogenesis in mice. BMC Complement Altern Med; 2002, 2: 4.

Bonamin LV, Nina AL, Caviglia F, Martinho K. Very low dilutions of dexamethasone inhibit their own pharmacological effect *in vivo*. Br Hom J; 2001, 90: 198–203.

Bronnikova TV, Schaffer WM, Olsen LF. Nonlinear dynamics of the peroxidase-oxidase reaction: I. Bistability and bursting at low enzyme concentration. J Phys Chem B; 2001, 105: 310–321.

Brown P, Meyer R, Cardone F, Pocchiari M. Ultra-high-pressure inactivation of prion infectivity in processed meat: A practical method to prevent human infection. Proc Natl Acad Sci USA.; 2003, 100(10): 6093–6097.

Calabrese EJ. Overcompensation stimulation: A mechanism for hormetic effects. Crit Rev Toxicol; 2001, 31: 425–470.

Calabrese EJ, Baldwin LA. U-shaped dose-responses in biology, toxicology and public health. Ann Rev Pub Health; 2001, 22: 15–23.

Delinick AN. A hypothesis on how homeopathic remedies work on the organism. The Berlin J Res Homoeopathy; 1991, 1: 249–253.

Elia V, Niccoli M. Thermodynamics of Extremely Diluted Aqueous Solutions. Ann N Y Acad Sci; 1999, 827: 241–248.

Elia V, Niccoli M. New physico-chemical properties of water induced by mechanical treatments. A calorimetric study at 25° C. J Therm Anal Calorimet; 2000, 61: 527–537.

Elia V, Elia L, Cacace P, Napoli E, Niccoli M, Savarese F. The "extremely diluted solutions" as multi-variable systems: a study of calorimetric and conductometric behaviour as a function of the parameter time. J Therm Anal Calorimet; 2006, 84 (2): 317–323.

Endler PC and Schulte J (Ed.).Ultra High Dilution Physiology and Physics. Dordrecht: Kluwer; 1994.

Endler PC, Pongratz W, Smith CW, Schulte J. Non-molecular information transfer from thyroxine to frogs with regard to homeopathic toxicology. Vet Hum Tox; 1995, 37: 259–260.

Hauser MJB, Olsen LF. The role of naturally occurring phenols in inducing oscillations in the peroxidase-oxidase reaction. Biochemistry; 1998, 37: 2458–2469.
Ho MW. The Rainbow and the Worm. The Physics of Organisms. Singapore: World Scientific; 2005.
Jonas W, Lin Y, Tortella F. Neuroprotection from glutamate toxicity with ultra-low dose glutamate. Neuroreport; 2001, 12: 335–339.
Liburdy RP. Electromagnetic fields and biomembranes. In: Ho MW, Popp FA, Warnke U (Eds). Bioelectrodynamics and Biocommunication. Singapore: World Scientific; 1994.
Luzar A and Chandler D. Hydrogen-bond Kinetics in liquid water. Nature; 1996, 379: 55–57.
Malarczyk E. How living organisms detect the small doses? A new approach to homeopathy. XXI Symposium of GIRI. March, 2007a; Monte Carlo, Monaco.
Malarczyk E. Kinetic changes in activity of HR-peroxidase, induced by very low doses of phenol. Int J High Dilution Res; 2008 (*in press*).
Malarczyk E, Pazdzioch-Czochra M. Multiple respiratory bursts as a response to veratrate stress in Rhodococcus cells. Cell Biol Intern; 2000, 24(8): 515–527.
Malarczyk E, Jarosz-Wilkołazka A, Kochma ska-Rdest J. Effects of low doses of guaiacol and ethanol on enzymatic activity of fungal cultures. Nonlinearity in Biology Toxicology and Medicine; 2003, 1(2): 176–184.
Malarczyk E, Kochmanska-Rdest J, Pazdzioch-Czochra M. Effect of low and very low doses of simple phenolics on plant peroxidase activity. Nonlinearity in Biology, Toxicology and Medicine; 2004, 2: 129–141.
Malarczyk E, Nowak G, Kochmanska-Rdest J. Changes in electrophoretic patterns of fungal oxizymes in the presence of very low doses of phenol and mediators. 4th International Medical Mushrooms Conference. 23–27 September, 2007b; Ljublijana, Slovenia.
Motwani JG, Fenwick MK, Morton JJ and Struthers AD. Furosemide-induced natriuresis is augmented by ultra-low-dose captopril but not by standard doses of captopril in chronic heart failure. Circulation; 1992, 86: 439–445.
Olsen LF, Hauser MJB, Kummer U. Mechanism of protection of peroxidase activity by oscillatory dynamics. Eur J Biochem; 2003, 270: 2796–2804.
Rey LR. Thermoluminescence of ultra-high dilutions of lithium chloride and sodium chloride. Physica; 2003, A323: 67–74.
Schaffer WM, Bronnikova TV, Olsen LF. Nonlinear dynamics of the peroxidase-oxidase reaction: II. Compatibility of an extended mechanistic model with previously reported model-data correspondences. J Phys Chem B; 2001, 105: 5331–5340.
Schulte J, Endler PC (Eds). Fundamental Research on Ultre High Dilution and Homeopathy. Dordrecht: Kluwer; 1998.
Tyihak E. Chapter 40 – In Floriculture, Ornamental and Plant Biotechnology. UK: Global Science Books, vol.III; 2007.
Tyihak E, Katay G, Manninger K, Steiner U, Schönbeck F. Non-linearity of plant resistance to pathogens induced by N-methylated compounds. Acta Phytopathol Enthomol Hungar; 2006, 41: 101–107.
Welch GR (Ed). Organized Multienzyme Systems. Orlando: Academic Press; 1995.
Welch GR, Clegg JS (Ed). The Organization of Cell Metabolism. New York: Plenum; 1987.

Chapter 10
The Effect of Homeopathically Prepared Thyroxin (10^{-30} Parts by Weight) on Highland Frogs: Influence of Electromagnetic Fields

Sabrina Weber, Peter Christian Endler[1], Sonja U. Welles, Erika Suanjak-Traidl, Waltraud Scherer-Pongratz, Michael Frass[2], Heinz Sprange, Gerhard Peithner[3], and Harald Lothaller[4]

Introduction

The amphibian model appears to be a useful tool for explaining certain phenomena encountered in homeopathy and homeopathy research (Endler et al., 1991, 1994a, 1995; Endler, 2003).

One particular aspect of the *similia principle*, iso-endopathy, can be demonstrated by first hyper stimulating *Rana temporaria* by immersion in an aqueous molecular thyroxin solution (10^{-8} parts by weight, not submitted to an agitation process) and then inducing an inverse effect by a homeopathically prepared solution (10^{-13}) of the same hormone thyroxin (Endler et al., 2003). There appears to be a relationship between the effect of homeopathically prepared thyroxin and a naturally or artificially elevated thyroxin level in the animals during metamorphosis. This is in some respects analogous to intoxication/detoxication studies, where organisms are first treated with a high dose of a toxin and then with a step-by-step diluted and agitated solution of the same toxin (Roth, 1991; Göldner, 2005). Significant curative effects have been observed, e.g. that of highly diluted Cd on Cd intoxicated frog spawn (Herkovits et al., 1993). Artificial speeding up of metamorphosis can not only be achieved by adding molecular thyroxin, but by enhancing temperature of the basin water. A curative effect of homeopathically prepared thyroxin on hyper stimulation caused by temperature may be discussed in the light of the homeopathic *similia principle*.

The *principle of potentisation* has been illustrated by the concurrent findings of researchers at the Zoological Institute of Graz University, the Department of Molecular

This article was first published in Homeopathy, 97:3–9, 2008 (Copyright Elsevier); with the same name and authors presented here

[1]Interuniversity College Graz – Castle of Seggau, Austria. college@inter-uni.net

[2]Medical University Vienna, Austria

[3]Peithner Inc., Vienna

[4]University of Graz, Austria

Presented at the XXIth Symposium of GIRI, Monaco, 2007

Cell Biology of Utrecht University, the Boltzmann Institute in Graz, the Zoological Institute of Vienna University and the Federal Institute for Veterinary Medical Investigation Graz. At all these institutions thyroxin solution (10^{-30}) led to a significant inhibition of metamorphosis speed (Endler et al., 1998; Zausner et al., 2002). The intervals between two successive points in time were *equal* within one and the same experiment, but varied from 8 to 72h depending on the overall duration of the experiment. For pooled data from these five laboratories (N = 1,620 per group), differences between groups were statistically significant with $p < 0.01$ (Chi-square-tests, Bonferoni-corrected) at all of the measuring points. Significance was also confirmed with logistic regression and multiple hazards model (Zausner et al., 2002).

Another characteristic, namely that of a *drug proving effect*, is illustrated by the finding that very frequent application of the solution, causes first a slowing down and subsequently an acceleration of metamorphosis in the test group. Here, the effect of the solution is at first inverse to and later concurrent with that of molecular hyperstimulation ("curative phase"/"drug proving") (Endler, 2003).

It goes without saying that this "frog's perspective" on homeopathy can be useful in demonstrating and illustrating certain phenomena related to homeopathy from the biologist's point of view, but that it does not cover the whole field of homeopathy in humans.

One characteristic of homeopathy research seems to be the difficulty of *inter researcher reproducibility* of experiments. For 17 years now, we have been engaged in finding, describing and varying the parameters that are crucial for the amphibian experiment (Welles et al., 2007). Initially we trained independent colleges at our own laboratory. As an interim stock-taking, we may state that our basic amphibian experiment (with highland tadpoles) tends to produce consistent results independent of whom they are performed by. Only recently, an independent college has successfully performed a similar experiment in Brazil (Guedes et al., 2004). Nevertheless, we have also started to do "research on the researcher". This line of investigation can be illuminating when people with "green thumbs" or "healing capacities" produce seemingly inexplicable results.

Information storage in the carrier substance is a crucial point in research on and practice of homeopathy. In the study presented here, we investigated the usability of the amphibian model to test the effect of various electromagnetic fields on a homeopathically prepared thyroxin solution. Starting with an early built microwave oven (the electromagnetic fields of which exceeded those generated by present-day microwave ovens), fields more commonly found in our technical environment were used, namely those of a mobile phone, an X-ray luggage inspection device at an airport and a red light barcode scanner (660nm) at the counter of a chemist's shop.

Our hypothesis was that the amphibian model could be useful to test such influences. With regard to the sources of the fields, we assumed that the early built microwave oven might be destructive to the homeopathically prepared solutions, with regard to the other devices, no hypotheses were formulated.

For control purposes and in continuation of our main line of research we also included repetitions of the basic experiment, i.e. with highland animals and no electromagnetic field.

Material and Methods

Researchers and Blinding

The experiments were carried out independently by Scherer, Suanjak and the team Weber/Welles in the laboratory of the Interuniversity College. All experiments, including application of test and control substances 10^{-30} as well as scoring of the stage of the animals, were performed blind. An external observer who came to the laboratory, the veterinary M. Wurm, was responsible for the blinding procedures. The same blinding method was used in each case. Substances 10^{-30} used for treatment (see below) were prepared in sets each consisting of the test solutions and the control solution. All substances were prepared in glass vials identifiable by the plaintext designation on the pull-off label. All solutions were left in their glass vials to avoid any extraneous influences through decanting. The plaintext labels were then removed by the blinding authority and replaced with labels bearing encoded designations. The code was not made known until after the presentation of the results. For reasons of laboratory convenience (danger of cross-contamination due to intricate handling) we abstained from using more than one vial per substance, all the more as the primary purpose of the study was to use the already existing model in order to learn about various influences on the test substance while leaving everything else unchanged. The project was organized by Endler.

Animals, Staging, Water and Further Laboratory Conditions

Rana temporaria larvae were taken from highland pools in the Austrian Alps at ca. 1,600 m above sea level. The starting stage was defined as the point at which the hind legs of the two-legged tadpoles are straddled such that one can only just see through the triangle formed by thigh, shank, and tail. This point of development occurs during Gosner's stage 31 (Gosner, 1960). The tadpoles were observed until the forelegs, which are preformed under the skin, broke through and the animals had thus entered the four-legged stage. In previous experiments with Scherer, different authorities from the Zoology Institute of Graz University as well as from the Environmental Agency of the County of Styria had carried out parallel counts to assess the counting method for reliability and obtained identical results. For the same purpose, one researcher had documented some of her counting results photographically.

Basins contained 8 l of dwell water each. Twenty animals were allotted to each of a number of white plastic basins according to a random procedure. This was done in the same way by each of the researchers/teams: 20–30 basins were used by each researcher/team. One by one, animals were fished out of the main tub and distributed over the basins so that there was one in each. This was repeated 20 times. The purpose of this procedure was to ensure that the animals were distributed homogenously in terms of their level of activity and swimming behavior in the main tub. The basins were arranged in rows of three to five according to the number of substances used in the experiment. The spatial arrangement of treatment groups within rows rotated from row to the next, i.e. basins with identical treatment groups were arranged in diagonals, and was left unchanged throughout the experiment to avoid the danger of cross-contamination through splashing. Indirect natural light was used. Room temperature was 21 ± 1°C. The tadpoles were fed with blanched greens (lettuce) ad libitum. Experiments were carried out in the laboratory of the Interuniversity College in September and October 2000–2005. Further details are given in Suanjak-Traidl (2005), Suanjak-Traidl et al. (2006), Welles (2006), Scherer-Pongratz et al. (2004) and Scherer-Pongratz et al. (2003).

Data Set and Experimental Design

A total of 2,980 animals was used, i.e. 149 basins with 20 animals in each of them. These formed a total of 21 groups according to treatment. Six different experiments were performed (rows of diagrams in Fig. 10.1, see below). In each of these experiments, one group (100–200 animals) was treated with control substance and one group (100–200 animals) with standard test solution (left column of diagrams in Fig. 10.1). Thus, 860 animals were treated with control solution and 860 with standard test solution. This part of the study can be understood as a modified repetition of the study describes above (Endler et al., 1991, 1995, 1998; Zausner et al., 2002), i.e. as mere fundamental research on a homeopathically prepared high solution of thyroxin. Due to the relatively high number of animals in each of the two groups, results should allow a distinct conclusion.

Furthermore, each of these six experiments had the purpose to investigate the influence of a certain type of electromagnetic fields on the test solution. This part of the study should be understood as pilot research on environmental factors that could influence the homeopathically prepared high solution of thyroxin. Due to the relatively small number of animals in each of the different groups, results will only allow to plan further study protocols.

Preparation and Administration of Hyperstimulation and Test Solutions: Expositure to Electromagnetic Fields and Shielding

First, all animals were exposed to the stock solution of tetra-iodo-thyronine sodium pentahydrate (T_4, Sigma, 10^{-4} parts by weight in double distilled water, diluted in the basin water down to a final concentration of 10^{-8}) (immersion in thyroxin 10^{-8}, hyperstimulation). In previous experiments with highland *Rana temporaria*, this has been found to speed up metamorphosis by around 5%. Depending on the experiment, one group was then treated with the homeopathically prepared standard

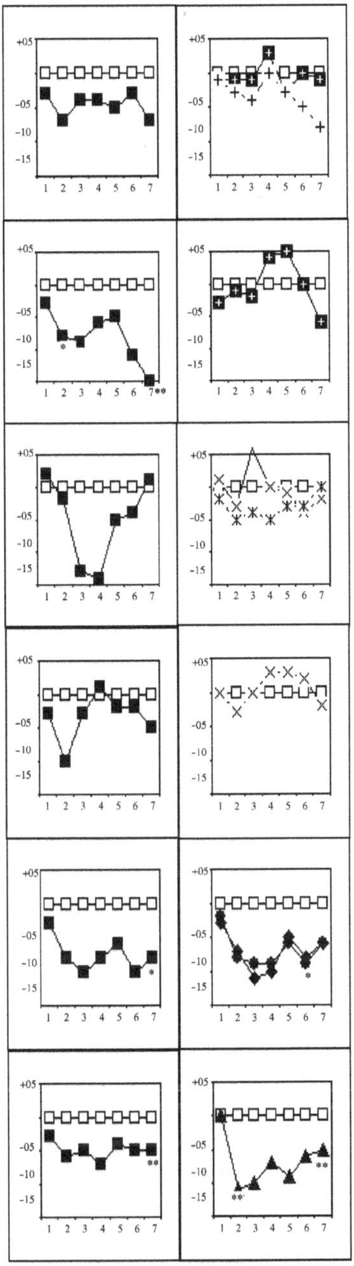

microwave, aqueous solutions

left:
white = WD30
(N = 100)
black = T D30
(N = 100)

right:
white squares = WD30
(N, see left)
white crosses = TD30 (microwave)
(N = 100)
black crosses = WD30 (microwave + protection) (N = 100)

microwave, ethanol solutions

left:
white = WD30
(N = 100)
black = T D30
(N = 100)

right:
white squares = WD30
(N, see left)
white crosses = TD30 (microwave)
(N = 100)

mobile phone, 0 cm

left:
white = WD30
(N = 100)
black = T D30
(N = 100)

right:
white squares = WD30
(N, see left)
X = TD30 (mobile)
(N = 100)
X! = TD30 (mobile + protection)
N = 100)

mobile phone, 25 cm

left:
white = WD30
(N = 160)
black = T D30
(N = 160)

right:
white squares = WD30
(N, see left)
X = TD30 (mobile)
(N = 160)

X-ray

left:
white = WD30
(N = 200)
black = T D30
(N = 200)

right:
white squares = WD30
(N, see left)
black rhombs = TD30 (X-ray)
(N = 100)
black stars = TD30 (X-ray + protection)
(N = 200)

red light scan

left:
white = WD30
(N = 200)
black = T D30
(N = 200)

right:
white squares = WD30
(N, see left)
black triangles = TD30 (scan)
(N = 200)

Fig. 10.1 The influence of highly diluted homeopathically prepared thyroxin, exposed to different electromagnetic fields. White squares = values of 4-legged control animals, normalized. Abscissa: differences of percentage of 4-legged animals; TD30 = agitated thyroxin 10^{-30}, WD30 = analogously prepared water. * = $p < 0.05$; ** = $p < 0.01$

test solution (thyroxin 10^{-30}, "TD30", see below), another one was treated with the analogously prepared solvent water (solvent 10^{-30}, "WD30") and, depending on the protocol, one or two groups were treated with the test solution that had been exposed to an electromagnetic field.

For preparation of the test solution thyroxin 10^{-30} the stock solution (10^{-4}) was further diluted with pure double distilled water in 26 steps of 1:10, and agitated after each step of solution according to standardized instructions. Using disposable pipettes, 1 ml of the precedent solution was added to 9 ml of water in a 20 ml glass vial. Agitation was performed by hand, with amplitude of ca. 20 cm: the vial was banged 30 times against a rubber impediment at intervals of approximately 0.5 s to create mechanical shocks. The solution thyroxin 10^{-30} ("TD30") is nominally 0-molar. For control, double distilled water was prepared analogously ("WD30").

Microwave oven, aqueous solutions: Measurements of electromagnetic fields in the immediate vicinity of the microwave oven (an early build) evidenced 50 Hz magnetic and electric fields of 110 μT and 5 V/m respectively and radiofrequency electromagnetic fields with an intensity of 2.45 GHz at 10 mW/m^2. The measured levels of 50 Hz magnetic flux density, generated by a transformer coil, and radiofrequency power density exceeded those generated by present-day microwave ovens. The microwave oven was used for irradiating two charges of thyroxin solution TD30. This was done placing the phials immediately next to the microwave oven and irradiating them for 100 s. One of these charges was left unprotected while the phial containing the other charge was wrapped in two layers of commercial aluminum foil. A third charge of TD30 and the water control WD30 were placed in a separate room and not exposed to the microwave oven.

Microwave oven, ethanol solutions: For this experiment, thyroxin charges and the water control were prepared with 42 vol % ethanol instead of water. One charge was exposed in the same manner as described above, one charge was not exposed. There was no charge wrapped in aluminum foil involved.

Mobil phone, 0 cm: Two charges of TD30 were exposed to a commercial type mobile phone with a near field equivalent, measured over a 100 m^2 sensor area, 300 microW/m^2 in standby and 339 miliW/m^2 during call set-up. This was done by placing the phials immediately next to the mobile phone and then performing five successive call set-ups with intermediate return to standby. One of these charges was left unprotected while the phial containing the other was wrapped in aluminum foil (Suanjak-Traidl, 2005; Suanjak-Traidl et al., 2006).

Mobile phone, 25 cm: One charge of TD30 was exposed to the mobile phone in the same manner as described above, the only difference being that it was placed at a distance of 25 cm from the mobile phone (Suanjak-Traidl et al., 2006; Welles, 2006).

X-ray: One charge of TD30 was exposed to an X-ray luggage inspection device at an airport during routine operation. The device had a nominal dose equivalent of 0.7 μSv (0.07 mrem or 0.7 mGy). Phials were exposed by passing them four times through the device at a conveyor speed of 0.2 m/s (Scherer-Pongratz et al., 2004).

Red light scan: One charge of TD30, bearing a paper label with a barcode, was exposed once to a red light barcode scanner (660 nm) at the counter of a chemist's shop (Scherer-Pongratz et al., 2003).

An estimate of how much of the applied electromagnetic energy possibly has been converted into thermal energy in the solutions leads to the conclusion that a raise of temperature can be neglected with regard to room temperatures that obviously have not influenced to test solution's activity in previous experiments.

3 μl of probe solution (WD30 or TD30) was added per animal and 300 ml of basin water at intervals of 48 h.

Comparison and Evaluation of Data

The cumulative frequency of animals treated with control or test substances having reached the 4-legged stage was aggregated for each day. Seven points in time were considered in each case to allow comparison between experiments. The interval between two successive points in time were equal within one and the same experiment, but varied from 1 to 3 days depending on the overall duration of the experiment (Welles et al., 2007; Suanjak-Traidl et al., 2006; Scherer-Pongratz et al., 2003, 2004).

Chi-square tests were used to compare groups. Different statistical methods had been discussed in connection with the amphibian model previously, such as variance analysis, t-test, survival analysis, proportional hazards model, logistic regression (Endler et al., 1991, 2003; Zausner et al., 2002); these usually lead to comparable results but: (1) need larger numbers of basins in one and the same experiment. Furthermore (2), depending on differences in the overall duration of experiments, S.D. is usually contorted when experiments from different laboratories are pooled. Interestingly (3), hyperstimulation (immersion in thyroxin 10^{-8}), as is used in the experiments described here, uses to lead to an increase of S.D. from about ± 1.0 to ± 1.5 for inert groups to about ± 2.5 to ± 3.0 for the hyper stimulated groups (1 S.D.) (Endler et al., 2003).

In order to make results comparable to those of previous publications, in the paper presented here, we restricted ourselves to the (comparatively rough and estimative) Chi-square test. Being aware of the problem of dependent data, however, we calculated P-values for each of the seven points in time that were used to describe every experiment.

Within each experiment, control solution was compared with standard test as well as with test solutions treated with the different electromagnetic fields. Results from standard test solution and control solution were pooled over all experiments; results from the test solutions treated with the different electromagnetic fields were not pooled over all experiments.

The absolute cumulative frequencies of four-legged animals have been published in different previous papers (Welles et al., 2007; Suanjak-Traidl et al., 2006; Scherer-Pongratz et al., 2003, 2004). In order to give an overview and to allow comparison between these findings, the present article deals only with frequency ratios. These have been depicted by normalizing the values of the control group to give a horizontal line in each chart which is arbitrarily defined as the 50% level (see Fig. 10.1). Each chart shows the difference between the % value of the control group and that of the relevant test group at seven points in time.

Results

860 animals were treated with homeopathically prepared thyroxin 10^{-30} (test group solution no. 1, standard test solution) and 860 animals were treated with standard control solution. In addition, further groups of animals (i.e. a total of 1,160 animals) were treated with solutions no. 2 and 3, respectively, according to the different protocols.

Control solution versus standard test solution: As can be seen in Fig. 10.1, left row, animals treated with the test solution (black squares) metamorphosed more slowly (5–10%) than the control animals (white squares). The number of test animals that reached the four-legged stage at defined points in time was smaller in the group treated with homeopathically prepared thyroxin at most of the points in time. This trend was found independently by all researchers involved in all six experiments. When experiments are considered separately, this trend, however, is statistically significant at a few points in time only (Fig. 10.1). When data from all six experiments are pooled, differences are statistically significant at some but not all (three out of seven) points in time: $p(2) < 0.05$, $p(3) < 0.05$, $p(7) < 0.01$. In other words, the effect of the homeopathically prepared thyroxin was opposed to the usual effect of molecular thyroxin. This result has been discussed in more detail in Welles et al. (2007).

Control solution versus test solution exposed to an electromagnetic field: In contrast (Fig. 10.1, right row), this effect did not occur when the thyroxin solution had been exposed to the fields of the early built microwave oven, or the mobile phone. There was no difference whether aqueous or alcoholic solutions were used, and there was, if any, only a small protective effect from aluminum foil. Practically no trends or statistical significances were found. On the other hand, airport X-raying and red light barcode scanning did not diminish the effect of the homeopathic solution. For result of statistic analysis of separate experiments, see Fig. 10.1. Data of test substances exposed to the different electromagnetic fields were not pooled.

Discussion

These data confirm that a special preparation process of stepwise solution and agitation (final concentration in the basin water 10^{-35} parts by weight) can lead to an inversion of the well-known stimulatory effect of thyroxin on amphibian metamorphosis.

Furthermore, these data suggest that the effect exerted by the thyroxin solution in the basic experiment is blocked when it is exposed to the field of an early built microwave oven, or to that of a commercial mobile phone. In the experiment where an aqueous and an alcoholic solution were compared, there was no difference with regard to their stability towards the fields of the microwave heater. There was, if any, only a small protective effect from the metal foil which some of the samples were wrapped in. On the other hand, in theses pilot experiments airport X-raying and red light barcode scanning did not diminish the effect of the homeopathic solution. Before these experiments have been subject to further in depth studies, it seems too early to try an explanation why some of the used fields do not, whereas the fields generated by other devices do affect the informational content of a homeopathic solution.

To our knowledge, however, the preliminary finding of a destructive influence of certain electromagnetic fields on homeopathic solutions is in agreement with the assumptions of many manufacturers and therapists as well as with the instructions they give to patients on how to store homeopathic preparations. Our findings suggest that great care is warranted in handling homeopathic preparations.

Further experiments with the amphibian model suggest that globules provide a more robust means of information storage. This notion finds support in established knowledge among homeopathy users. One must consider that most homeopathic remedies on the market today are sold in the form of globules rather than of solutions. Experiments with globules are on-going.

Biophysical theories have evolved which support the possibility of such findings. Physics research has revealed that water dipoles may develop phase coherent oscillations through radiation coupling (Del Giudice and Preparata, 1998). It was proposed that these could be modulated as a time-ordered pattern of signals. The present authors believe that the theoretical explanation of homeopathy – similarly to the explanation of a wide range of other phenomena in physiology, psychology and epistemology – could be broadened by the application of de Broglie's concept of the wave nature of particles and the particle nature of waves (Schulte, 1998). Biophysical questions have been addressed with the help of the amphibian model (Endler et al., 1994).

A possible methodological refinement for further studies would be to use *un*succussed solvent as an additional control, apart from the succussed solvent used in the experiments presented here. Furthermore, blank solutions exposed to the electromagnetic fields should serve as additional controls.

Summing up a side finding presented in the results section of this paper and discussed in (Welles et al., 2007), further experiments on *highland* amphibian should be performed without pre-treatment by molecular thyroxin as, in contrast to our previous assumption derived from *lowland* amphibian (Endler et al., 2003; Graunke et al., 2007), such pre-treatment does not seem to enhance the effect of the homeopathically prepared thyroxin solution 10^{-30} in highland animals. It may be hypothesized, however, that the increase in water temperature from the highland biotope (e.g. 15°C) to laboratory room temperature (21 ± 1°C) is an important stimulus that makes highland amphibian sensitive to homeopathically prepared thyroxin. This remains to be further investigated.

It goes without saying that, with differences of 5–10% between groups, results with the amphibian model are statistically significant only when an adequate number of animals are included in the study, i.e. observed effects are modest with regard to patient's expectations.

A comprehensive overview on basic questions and problems and the state of the art in research on homeopathy has been given in (Endler, 2003).

Acknowledgements Thanks are due to Peithner Inc., Vienna, for their expertise and help in electronic and technical issues. Thanks are due to the editor of *Homeopathy* for his suggestion that our team should compile main findings and publications on the amphibian model since 1991 when the model was first presented in that journal. Last not least, thanks are due to *Homeopathy* (Elsevier) to allow reprint of this article exclusively in this issue.

References

Del Giudice E ,Preparata G. Coherent electrodynamics in water. In: Schulte J, Endler PC (Eds). Fundamental Research in Ultra High Solution and Homoeopathy. Dordrecht: Kluwer; 1998.

Endler PC. Homeopathy – An Expedition Report. An Old Healing System Gains Plausibility. Graz: edition@inter-uni.net; 2003 (ISBN 3–9501448–1–1).

Endler PC, Pongratz W, Kastberger G et al. Climbing activity in frogs and the effect of highly diluted succussed thyroxine. Br Hom J; 1991, 80: 194–200.

Endler PC, Pongratz W, Kastberger G, Wiegant FAC, Schulte J. The effect of highly diluted agitated thyroxin on the climbing activity of frogs. Vet Hum Tox; 1994a, 36: 56–59.

Endler PC, Pongratz W, van Wijk R et al. Transmission of hormone information by non molecular means. FASEB J; 1994b, 8 (4): A400 (Abs.).

Endler PC, Pongratz W, Smith CW, Schulte J. Non-molecular information transfer from thyroxin to frogs with regard to homeopathic toxicology. Vet Hum Tox; 1995, 37: 259–260.

Endler PC, Heckmann C, Lauppert E et al. The metamorphosis of amphibians and information of thyroxine. In: Schulte J, Endler PC (Eds). Fundamental Research in Ultra High Dilution and Homoeopathy. Dordrecht: Kluwer; 1998.

Endler PC, Lüdtke R, Heckmann C et al. Pretreatment with thyroxin (10^{-8} parts by weight) enhances a "curative" effect of homeopathically prepared thyroxin (10^{-13}) on lowland frogs. Res Compl Med; 2003, 10: 137–142.

Göldner C. Review und Evaluierung von neueren, mit homöopathischen Zubereitungen durchgeführten toxikologischen Experimenten. Austria: Dissertation: Graz University 2005.

Gosner KL. A simplified table for staging anuran embryos and larvae with notes on identification. Herpetologica; 1960, 16: 183–195.

Graunke H, Scherer-Pongratz W, Frass M, Endler PC. Treatment of lowland frogs from the spawn stage on with homeopathically prepared thyroxin (10^{-30}). Accepted by The Scientific World Journal, 2007.

Guedes JR, Ferreira CM, Guimaraes HM et al. Homeopathically prepared solution of *Rana catesbeiana* thyroid glands modifies its rate of metamorphosis. Homeopathy; 2004, 93(3): 132–137.

Herkovits J, Perez-Coll C, Zeni W. Reduced toxic effect of Cd on bufo arenarum embryos by means of very high diluted and stirred solutions of Cd. Communicationes Biologicas; 1993, 7: 70–73.

Roth C. Literature review and critical analysis on the topic of in- and detoxication experiments in homoeopathy. Berlin J Res Hom; 1991, 1: 111–117.

Scherer-Pongratz W, Endler PC, Haidvogl M et al. Das Scannen homöopathischer Arzneimittel. Untersuchung am zoologischen Modell. Österreichische Apotheker-Zeitung; 2003, 8: 379–381.

Scherer-Pongratz W, Endler PC, Haidvogl M et al. Röntgenkontrolle homöopathischer Arzneimittel. Untersuchung am zoologischen Modell. Österreichische Apotheker-Zeitung; 2004, 12: 564–565.

Schulte J. Bio-Information between quantum and continuum physics. In: Schulte J, Endler PC. (Eds). Fundamental Research in Ultra High Solution and Homoeopathy. Dordrecht: Kluwer; 1998.

Suanjak-Traidl E. Homeopathically prepared thyroxin (10e–30) and the metamorphosis of hyperstimulated highland *Rana temporaria* (I). MSc dissertation: college@inter-uni.net, Graz; Austria, 2005.

Suanjak-Traidl E, Weber S, Welles SU et al. Handy-Exposition homöopathischer Arzneimittel – Untersuchung am zoologischen Modell. Österreichische Apothekerzeitung; 2006, 20: 973–975.

Welles SU. Homeopathically prepared thyroxin (10e–30) and the metamorphosis of hyperstimulated highland *Rana temporaria* (II). MSc dissertation: college@inter-uni.net, Graz, Austria, 2006.

Welles SU, Suanjak-Traidl E, Weber S et al. Does pretreatment with thyroxin (10^{-8} M) enhance a "curative" effect of homeopathically prepared thyroxin (10^{-30}) on highland frogs? Results of a multi researcher study. Accepted by Res Compl Med, 2007.

Zausner C, Lassnig H, Endler PC et al. Die Wirkung von "homöopathisch" zubereitetem Thyroxin auf die Metamorphose von Hochlandamphibien – Ergebnisse einer multizentrischen Kontrollstudie. Perfusion; 2002, 17: 268–276.

Part IV
Clinical Research

Chapter 11
Clinical Verification of Homeopathic Symptoms: Experience and Statistics

Lex Rutten[1] and Michel van Wassenhoven[2]

Introduction

Successful homeopathic prescriptions are not based on diagnosis only. The population with a certain diagnosis or disease is divided in subgroups, each subgroup responding to another homeopathic medicine. The necessity to divide 'diagnosis-populations' into subgroups is also becoming relevant in conventional medicine and as such studied in pharmacogenetics (Swen et al., 2007). Homeopathic doctors use symptoms (personal traits) to recognize subgroups in the same way as pharmacogenetic tests. Clinical and pharmacogenetic tests render probabilities; the probability of a diagnosis is sequentially updated after each test that indicates the diagnosis. This updating process is effectuated in Bayesian statistics. It is possible to describe the homeopathic prescription as a Bayesian algorithm.

Verification of homeopathic symptoms has so far been a neglected field. What is the use of efficacy research if the instruments of the method have serious flaws? The most serious flaw of the homeopathic repertory is that entries of medicines in a symptom-rubric are based on absolute occurrence of the symptom in the cured population, not on prevalence. This way, frequently used medicines are over-rated, especially regarding frequently used symptoms. Bayesian theory shows that the prevalence of the symptom is the only correct criterion. A symptom is an indication for a specific medicine only if the prevalence of that symptom is higher in the population cured by that medicine than in the rest-population.

If one keeps adding medicines to the repertory-rubrics and materia medica using the old rule the computer could become a serious threat to the instruments' quality. Sooner or later, each symptom will turn up in the population that is cured by a specific medicine, by mere chance, even if it is seldom seen in that medicine-population. The computer makes it very easy to update the content of the repertory. It has become a sales-argument for electronic repertory manufacturers to have the largest

[1] Dutch Association of Homeopathic Physicians, Breda, Netherlands. lexrtn@concepts.nl

[2] European Committee for Homeopathy, Brussels, Belgium

Presented part at the XVIIIth Symposium of GIRI, Brussels, 2004 and part at the XIXth Symposium of GIRI, Monaco, 2005

repertory with the most entries and their income depends on updates. As yet there are no clear and commonly used quality criteria for additions to the repertory.

The entries in homeopathic repertory are based on expert opinion, mostly of very few experts. Moreover, their experience is seldom systematically collected and mostly memory-based. Even experienced homeopathic physicians have a limited number of successful cases concerning one medicine. Symptoms that are daily used, 'keynotes', are relatively infrequently occurring, but also not too seldom. The estimates of occurrences of most keynotes lie between 5% and 10%. These facts cause mistakes based on statistical variance. Systematic recording of symptoms and multi-centered gathering of data could overcome these problems (Rutten et al., 2006).

Bayesian statistics enable us to assess homeopathic symptoms (Stolper et al., 2002), but there are also other modern statistical tools to constitute the homeopathic medicine picture in a scientific way, like Canonical Discriminant Analysis (CDA).

In this chapter, some possible ways to assess homeopathic symptoms with scientific instruments are shown. First, Bayesian statistics is introduced. In the methods section, the various methods used in three projects are presented, the first is long-term recording of all *repertorisations* and prescriptions in one practice and its retrospective analysis, the second is retrospective evaluation of successful cases of a group of Dutch doctors by consensus meetings, and the third is multi-centered prospective assessment of six homeopathic symptoms. Some aspects of the discussion are placed under the results of the respective methods; the more general aspects of the discussion are placed after the results.

Bayesian Statistics

Experience from the past indicates success in the future. If 'many' patients that respond well to the homeopathic medicine *Lachesis* are loquacious, loquacious patients will more likely respond well to *Lachesis*. Herein, 'many' means 'more than in the rest-population', the rest-population being the population that responds well to other medicines. Here lies the difference with the existing criteria to enter a medicine in a repertory-rubric. In 1763 reverent Thomas Bayes published his theorem describing the way we learn from experience (Bayes, 1763).

From Experience to Prognosis

The Bayesian principle is in fact quite simple: a diagnostic test is better as it is positive more frequently in people with the disease than in other people. Hahnemann also made this observation and in the same fashion he concluded that rare and peculiar symptoms are the most valuable symptoms (Table 11.1). Likelihood ratio (LR) expresses the relation between the prevalence of the symptom in the population with the illness and the population without the illness. In the homeopathic translation:

Table 11.1 Two aphorisms by Hahnemann translated into Bayesian Likelihood Ratio

	Rhus-t.	Others	LR
"Getting better by motion is 10 times more frequent in Rhus-t. patients than in other patients"	50%	5%	**10**
"Rare symptoms are most important"	20%	0.1%	**200**

the relation between the prevalence of the symptom in the population cured by a certain medicine and the rest of the population.

A symptom with a higher LR is more important. Peculiar symptoms have high LR because they are specific for just a few medicines (Rutten, 2007). The Bayesian formula comes in different variants; one of the simple and intuitively understandable forms is as follows:

Posterior odds = LR x prior odds[1]

If a symptom is as frequent in the population cured by a certain medicine as in the rest-population LR = 1. Such a symptom gives no indication at all for this medicine. When a medicine is frequently prescribed, like Phosphorus, a number of patients that take this medicine and are thistles will be seen. But when this occurs as frequently as in the rest-population, LR = 1 and the symptom is no indication for Phosphorus.

Homeopathic Diagnosis

The choice of a homeopathic medicine is usually not based on one fact (diagnosis). In Bayesian perspective we can describe the decision-process of a homeopathic physician: if we add symptoms, our certainty about the curative effect of a medicine will grow; if our symptoms are better (e.g. if they are peculiar) our certainty will grow faster. Suppose that the chance that a homeopathic medicine cures is 1% if there are no symptoms, than our conviction that *Rhus toxicodendron* will be curative grows as in Table 11.2 with three subsequent symptoms (in this example odds are translated into chance by the computer).

This is a normal procedure in homeopathy. The patient visits the doctor because of joint pains. The homeopathic doctor then asks about circumstances that influence the complaint. If the patient tells him that the pain is ameliorated by motion, the homeopathic doctor thinks of *Rhus toxicodendron* as one of the possibilities. If further investigation learns that the patient has a definite desire for cold milk his expectation that *Rhus toxicodendron* will help grows. The last symptom, aggravation from wet weather, is subsequently enough to prescribe this medicine. This is a rather simplistic description of the decision algorithm in homeopathy and supposes mutual independence of the various symptoms. If one assesses more than one symptom in one investigation, independence of these symptoms should be checked.

[1]Odds = chance/(1-chance); in words: the chance that something will happen divided by the chance that it will not happen. Odds = 1 means: chance is fifty-fifty.

Table 11.2 Hypothetical decision process to estimate chances of cure by the homeopathic medicine *Rhus toxicodendron*

	Symptom	LR	Chance of cure (cumulative)
0	No symptoms		1%
1	Motion ameliorates	10	9%
2	Desire cold milk	6	37%
3	Wet weather aggravates	4	72%

Methods

More information about the prevalence of symptoms (instead of absolute occurrence) can be obtained by systematic gathering of clinical data. Since 1997, three projects are been performed in clinical practice.

Two of the projects are related, the Materia Medica Validation (consensus procedure) and the prospective LR assessment were both initiated by the Committee for Methods and Validation (CMV) of the Dutch association of homeopathic physicians (VHAN). The prospective LR assessment was a result of evaluation of the Materia Medica Validation. Part of the doctors participating in the Materia Medica Validation also participated in the LR project. The retrospective LR assessment is linked to the Cli-Fi-Col project which is briefly described below. The results of the three projects are only partly comparable. To make some comparison between the three projects, one focused on two symptoms assessed in the prospective LR-project, the vague symptom 'Sensitivity to injustice' and the rather concrete symptom 'Recurrent herpes lips' to indicate possibilities and limitations of the method. A small number of medicines were highlighted for this purpose.

Practice Registration

Since the eighties, many homeopathic MD are using individual computers for help to homeopathic medicine selection and clinical data collection. As such many private clinical data files exist. Retrospective studies are possible and centralization of data can be organized. One example of such a data file is presented. The Cli-Fi-Col (Clinical File Collection) has just started at an international level (Fig. 11.1). The databank is localized in Italy and is able to receive, on line, all clinical files from all medical doctors that are using the Winchip® program. Cases follow-up and prescription's results are automatically updated day by day.

Since 2002, a pilot open cohort study on one private practice has been performed and retrospectively aimed at evaluation of the use of LR statistics on a clinical file in daily practice. Individual records were based on *repertorisations*, only symptoms in the repertory program (RADAR®) were recorded. A potential

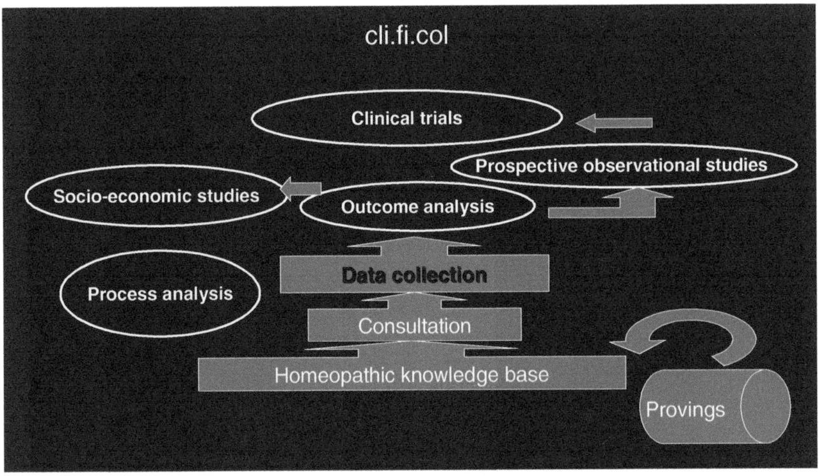

Fig. 11.1 Organization of the Cli-Fi-Col project

source of bias is that the repertory has a number of rubrics which are semantically similar, but contain different medicines. Symptom-rubrics in the repertory are not clearly defined.

These results are confirming some homeopathic basic knowledge used to prescribe homeopathic medicine to patients and questioning others. The validity of the study depends only of the amount of complete clinical files introduced in the databank. Results of treatments must be evaluated in a systematic way following a comparable scale.

Data collection is the centre of clinical verification of homeopathic medicine. Using a centralized databank, prospective clinical studies of all kinds are possible.

For each medicine and each symptom used has a decision key for the medicine prescription LR value has been calculated. The positive gradient (LR+) is significant when it is upper than 1. It is an indication that the medicine has been effective in presence of this symptom. Higher above 1, better it is, but one must take into account the confidence interval, if it is too broad, no final conclusion is possible.

This study is a retrospective study. It considers the symptoms prevalence in a patient's databank. This databank is considered being representative of people using homeopathy. This analysis of a databank is a fortiori limited since it is performed looking at results of only one physician. Nevertheless the advantage of this situation is its coherence and so the interpretation of results is easier.

Materia Medica Validation

Since 1997 the Committee for Methods and Validation of the Dutch homeopathic doctors association organizes consensus meetings regarding homeopathic medicines.

Participants (usually 12–25) present their selected best cases concerning two scheduled homeopathic medicines. Each case is assessed by a group of colleagues concerning the probability of causal relation between medicine and cure and the 'completeness' of the cure. For further evaluation only cases with apparent causal relation between medicine and effect, effect according to degree 3 or 4 on the GHHOS scale (see later) and curative effect longer than one year are accepted.

Retrospectively, the prevalence of symptoms is assessed. Participants are also asked to give their expert opinion about the prevalence of the same symptoms in their practice populations. This way estimates are made about Likelihood Ratios of homeopathic symptoms. The presented cases were mostly old cases where it was impossible to retrieve all symptoms. Only the most well-known symptoms of the medicine were checked systematically.

Prospective Assessment of Six Homeopathic Symptoms

Since June 2004, in an open cohort study, 10–12 Dutch practices investigate the (lifetime) prevalence of six homeopathic symptoms in consecutive new patients and record results on the prescribed medicines according to a specification of the Glasgow Homeopathic Scale (GHHOS). The symptoms are: (1) 'Diarrhea from anticipation', (2) 'Fear of death', (3) 'Grinding teeth during sleep', (4) 'Recurrent herpes lips', (5) 'Sensitive to injustice' and (6) 'Loquacity'. This selection was made to test a variety of vague and concrete symptoms. The outcome was analyzed in two ways:

1. Likelihood ratio (LR) assessment is a straightforward comparison of the prevalence of the symptom in the population cured by a certain medicine with the rest-population.
2. Multivariate analysis: correlation matrices, Principal Component Analysis (PCA) and Canonical Discriminant Analysis (CDA). Correlation matrices and PCA check for inter-relatedness of symptoms. CDA is a statistical analysis, rendering 'types' of patients. The maximum number of types is restricted by the number of assessed symptoms (maximum six types) and the number of medicines (maximum number of medicines minus one).

First the symptoms were specified during a consensus meeting; e.g. the number of recurrences of herpes lips and criteria for 'sensitivity to injustice'. The six symptoms chosen are known as so-called keynotes for different medicines. This choice prevents interrelation of the symptoms. The project ended December 2007. All consecutive patients older than two years were included. All six symptoms were checked in all included patients. All prescribed medicines were recorded and results related to each medicine were monitored during the treatment using the GHHOS scale, see also under 'Discussion - Gold standard'. Results GHHOS 3 and 4 were only recorded after at least 6 months of follow-up for that medicine. Results were adjusted when changed, but after one year follow-up of one medicine, the result was fixed. Participating doctors receive feedback on their results during consensus meetings twice a year. Data were entered in different computer programs for practice management, but the exported data were all in the same format.

For clarification of methods, one focuses on the symptoms 'Sensitive to injustice' and 'Herpes lips'. The results can be compared with rather small rubrics in the existing repertories, see also under 'Discussion – Vagueness'.

'Sensitivity to injustice' was defined as: sensitive to injustice inflicted to others, resulting in actions like switching off the TV on reports about Darfur, or writing letters to politicians and newspapers about social injustice, etc. There are several different versions of the symptom-rubric 'Sensitive to injustice'. For this text, RADAR-Synthesis® version 8.1.40, 'modern to 1987' was used, containing the following medicines: Calc (Calcarea carbonica), **Caust (Causticum)**, Cupr (Cuprum), Dros (Drosera), *Ign (Ignatia),* Merc (Mercurius), Nux-v (Nux-vomica), Sep (Sepia), **Staph (Staphisagria)**, Verat (Veratrum album).

The entry of Ignatia in italics indicates that the symptom is a stronger indication for Ignatia than for the medicines in plain type. Entries in bold type indicate that the presence of the symptom is the strongest indication for those medicines.

The symptom 'Sensitive to injustice' was not mentioned in Kent's original repertory and also not in some earlier versions of RADAR. Some doctors chose the rubric 'Anger with indignation' for this symptom. This rubric does not contain *Causticum.*

Sensitivity to injustice is a very vague, socially desirable and subjective symptom and as a contrast one also assessed the symptom 'Herpes lips'. 'Herpes lips' was defined as more than six times a year, not necessarily shortly before treatment, but during a considerable period in life. The rubric of the same RADAR version contains 44 medicines. The prevalence of this symptom was compared with existing literature (Shulman et al., 1992; Axell and Liedholm, 1990; Katz et al., 2001; Embil et al., 1975).

For the evaluation of the prevalence of symptoms in the whole population, all records were used. For the evaluation of effective prescriptions, patients that were lost to follow-up were not taken into account. For calculation of LR the prevalence of the symptom in the population responding to the specific medicine with result GHHOS 2–4 (medicine population) was compared with the prevalence in the whole population minus the medicine population, see also under the discussion. GHHOS 2 result means that not only the presented complaint was improved, indicating a 'homeopathic' effect. LR's were calculated in Excel spreadsheets. For relevant LRs 95% confidence intervals were calculated using the program CIA (BMJ) (logarithmic method). Pivot tables were inspected visually for possibly interrelated medicines. 300 Possible correlations were tested in correlation matrices.

For Principal Component Analysis (PCA) the population responding well to 20 medicines – that were most frequently prescribed – was chosen. These 20 medicines are responsible for about 900 out of 1,700 successful prescription.

For Canonical Discriminant Analysis (CDA) 10 medicines that were frequently prescribed were chosen and had a relation with the symptoms according to LR results. The discriminant analysis was not planned before the project, in that case one should have included more symptoms. The resulting six types of patients, restricted by the number of symptoms, are just an indication of the possibilities of CDA of this kind of data. Multivariate analyses (correlation matrices, Principal Component Analysis and Canonical Discriminant Analysis) were performed in SPSS.

Multivariate Analysis

Computers enable us to analyse huge amounts of data. One of the possibilities is pattern recognition. An example: Satellite pictures consist of pixels in various densities. These densities vary according to the kinds of soil or vegetation. The pixel-density of each vegetation has a certain mean and a certain intra-group variance specific for that vegetation. Another vegetation has another mean pixel-density and the difference between the pixel-densities of two vegetations is called the inter-group variance. By calculating the difference between intra-group variance and inter-group variance of parts of the picture the computer can tell which kind of vegetation is in a certain part of the satellite image. This is done by Canonical Discriminant Analysis (CDA). In homeopathy, a specific symptom has different 'densities' in various populations cured by different medicines. One can use CDA to show which symptoms correlate best with which medicines.

Principal Component Analysis (PCA) is used to discover inter-relatedness of symptoms. Comparing abilities of schoolchildren, like geometry, arithmetic and algebra, one will see much correlation because these are all mathematical abilities. Three abilities can be summarized in one denominator. Such relations can be detected by correlation matrices and PCA, correlation matrices for two-factorial relations and PCA for multi-factorial relations.

Results Practice Registration

The data bank gathered 3,538 patients. 21,327 consultations are available for this study (2,148 patients). Patients for which prescriptions result was not yet encoded (recent patients) were eliminated as also the consultations, which led to several prescriptions (several different medicines at the same time). For these cases, it is not possible to give a treatment result linked to a single medicine. As an example, the results for *Staphisagria* are shown.

Symptom Verification of Staphisagria

For *Staphisagria*, 266 various symptoms from 28 chapters of the repertory were used for this prescription among 58 patients. The values LR + et LR- being significant only if a sufficient number of patients is included in the study; first the symptoms were suppressed leading to the prescription for only one or two patients and also the results LR + lower than 1. So, 25 symptoms remain analyzable; for five, the values are already really significant.

For five symptoms, the results are positively significant:

Mind – Ailments from anger with indignation: is a pathogenetic symptom, the link between this symptom and the efficiency of the medicine is evident. The importance of this symptom is confirmed by a significant LR- value.

Mind – Ailments from disappointed love: is a pathogenetic symptom, the link between this symptom and the efficiency of the medicine is evident.

Mind – Ailments from anger: is a pathogenetic symptom, the link between this symptom and the efficiency of the medicine is evident. The importance of this symptom is confirmed by a significant LR- value.

Mind – Anxiety of conscience: is a pathogenetic symptom, the link between this symptom and the efficiency of the medicine is evident and here also the absence of this symptom could be an exclusion criterion. The value of STAPH in this Synthesis rubric should be revalued.

Perspiration –Profuse: is a pathogenetic symptom, the link between this symptom and the efficiency of the medicine is more than probable.

One probable symptom could be added: *Skin – Unhealthy* – it is a pathogenetic symptom.

> For these six symptoms, the link between the presence of the symptom (and several times his absence) and the efficiency of the medicine is clear. All these symptoms were described in the proving and form a part of the most known characteristics of this medicine. As such the similia and the globality law are also verified.

For 19 symptoms, the results must be confirmed by further studies (by category of decreasing validity):

1. Mind – Ailments from anger with silent grief: is a pathogenetic symptom.
2. Mind – Concentration, difficult: is a pathogenetic symptom.
3. Mind – Anger, violent: is a pathogenetic symptom.
4. Mind – Anger: is a pathogenetic symptom.
5. Mind – Discontented with himself: is a pathogenetic symptom.
6. Mind – Contradiction, intolerant of: is a pathogenetic symptom.
7. Mind – Ailments from mortification: is a pathogenetic symptom.
8. Mind – Fear of impending disease: is no pathogenetic symptom.
9. Mind – Ailments from grief: is a pathogenetic symptom.
10. Mind – Aversion of company: is a pathogenetic symptom.
11. Mind – Fear something will happen: is no pathogenetic symptom.
12. Head – Perspiration – Scalp: is no pathogenetic symptom.
13. Mind – Confidence, want of self: is not really a pathogenetic symptom.
14. Mind – Ailments from anticipation: is not really a pathogenetic symptom.

15. Mind – Restlessness: is a pathogenetic symptom but little specific.
16. Mind – Anguish: is no pathogenetic symptom, STAPH is not in this Synthesis rubric.
17. Mind – Mood changeable: is a pathogenetic symptom.
18. Sleep – Sleeplessness from activity of thoughts: is a pathogenetic symptom.
19. Mind – Irritability: Very important rubric less specific. Nevertheless it is a pathogenetic symptom.

Up to now, looking at the results in attached table, only pathogenetic symptoms are verified for this medicine. A lot of authors pointed out the importance of STAPH in non verified rubrics. Therefore, homeopathic repertories are not completely evidence based and values of medicines in several rubrics (symptoms) would be reconsidered using evidence based clinical results of treatments. It will allow a more efficient use of homeopathic repertories in the future.

The similarity and global law associates a strong causality (result of anger, the absence of this item could exclude the prescription of this medicine) to the diseased symptoms, certainly if the clinical expression is dermatological. This is traditional homeopathic knowledge, not only the symptoms but also these associations of symptoms are today verified with this study.

About hundred medicines and thousands of symptoms are able to be analyzed. Complete results of this statistical analysis will be published in a specific book.

Results: Materia Medica Validation

Bayes' theorem and our data after bringing cases from several doctors together can also explain the problem with expert opinion. During *Materia Medica Validation,* 10–25 experienced doctors bring their best cases concerning one homeopathic medicine. It appears that even experienced doctors seldom have more than three 'best' cases concerning one medicine. The participants knew nothing about the results of practice registration mentioned before. In November 2002, the consensus meeting (retrospectively) showed two out of 18 *Staphisagria* patients (11%) who were sensitive to injustice. The prevalence of the symptom 'sensitive to injustice' in the whole population was estimated to be 10%. Later (February 2007), the prospective LR research confirmed that the prevalence of the symptom was indeed 10% in 3,367 patients, and two of the 23 patients (8%) that responded well to *Staphisagria* during the LR research were sensitive to injustice (see later).

Many entries in the repertory are based on the opinion of one expert (Rutten et al., 2004b). Suppose that the editor of the repertory asks the opinion of an experienced homeopath with five good *Staphisagria* cases. Exact binomial chance calculation based on the data of our prospective research indicates that the chance that one of these five patients is sensitive to injustice is about 47%. So one out of two doctors with five cases estimates that the prevalence of the symptom in the *Staphisagria* population is one out of five (20%), being more than his estimate of one out of 10 (10%) for his whole practice. So in the experience of this doctor, from his *Staphisagria* population, the symptom 'sensitive to injustice' is an indication for *Staphisagria*. There is even a possibility of 11% (corresponding with one out of nine doctors) that two out of five *Staphisagria* cases (40%) are sensitive to injustice, meaning that 'sensitive to injustice' is a rather strong indication for *Staphisagria*. The gathering of cases of 14 experienced colleagues in our consensus meeting and LR assessment increased numbers and learned that the real prevalence of this symptom in the *Staphisagria* population is less. 'Sensitivity to injustice' might be no indication for *Staphisagria*.

MM validation is retrospective analysis of cases long ago. Not all symptoms are checked in every patient contrary to prospective LR investigation the symptom is checked in every patient. But, in prospective investigation, the prevalence of symptoms might be exaggerated. Sensitivity to injustice is rather subjective and a desirable character trait. But it is possible to compare the population that responded well to *Staphisagria* with the rest-population.

MM Validation of Staphisagria

Table 11.3 shows part of the validation outcome of 18 *Staphisagria* cases. Not all symptoms mentioned by the participants are recorded in this table. Many symptoms were mentioned once, like headache, fatigue, aphthae, flatulence, diarrhea. The estimation is that headache, etc. occurs in more than 10% of all patients. The prevalence of headache in *Staphisagria* population was therefore not greater than in the rest-population. According to Bayes' law this means that migraine is no indication for

Table 11.3 MM validation of *Staphisagria*

Total number of patients = 18 Symptom	n=
Anger	17
Suppressed anger	10
Throws things	2
Easily hurt	4
Ailments from mortification	3
Sensitive to injustice	2
Never complaining	3
Fear during abdominal pain	1
Early caries teeth	2
Looseness teeth	1
Pressure in root of teeth	1
Sensation anus dry	1
Chronic conjunctivitis	2
Headache looking at revolving objects	1
Scabs skin: ichtyosis, psoriasis	2
Pain hollow of knee < morning	1

Staphisagria. It doesn't contraindicate this medicine either. The selection in this table consists of symptoms that could be indications for *Staphisagria* according to our estimation because their prevalence seems higher than in the rest-population. For a symptom like 'Fear during abdominal pain' one patient suffices, because one estimates that the prevalence of this symptom in the rest-population is lower. Frequently, occurring symptoms need more confirmation than peculiar symptoms. Therefore, Hahnemann's aphorism 153 is a specification of Bayes' theorem.

Most patients have anger, but this could be due to confirmation bias. This result is consistent with the practice registration. The existing materia medica relates *Staphisagria* strongly to (suppressed) anger. Possible *Staphisagria* is not prescribed if the patient has no anger. The symptom 'Sensitive to injustice' is recorded twice in this population, but since the prevalence of this symptom is not known in the rest-population, no inferences can be made about the importance of this symptom regarding this medicine.

Results of MM validation and LR assessment demonstrate that many entries in the repertory could be influenced by chance. Bias, like confirmation bias, could also influence expert opinion and the prescription of medicines. This cannot be ruled out by MM validation or LR assessment. The symptom 'Recurrent herpes lips' was not detected in this population, possibly because of recall bias – this symptom is not expected in relation to this medicine (see later). This kind of bias can be ruled out by prospective investigation.

Results of Prospective Study

The prospective study started June 2004. After 33 months 3,367 patients entered the study and 3,246 prescriptions were evaluated. The symptom 'Sensitive to injustice' was present in 330 patients (10%). This prospective study enables us to compare medicine populations with the rest-population.

The difference between absolute occurrence and prevalence is essential. The fact that a symptom is repeatedly observed (absolute occurrence) in a population improved by a certain medicine was so far the criterion to include the medicine with emphasis in the repertory-rubric, but is not enough to apply Bayes' theorem. If the symptom is not rare and if the medicine is often used, like *Calcarea*, it is possible that the symptom is not characteristic for *Calcarea* patients as a group.

For this first prospective assessment of homeopathic symptoms, vagueness was also under interest. Beside vague symptoms like 'Sensitive to injustice', rather concrete symptoms were also considered like 'Herpes lips'. The results for the symptom 'Herpes lips' are in Table 11.4. Only results with more than one patient with herpes lips are represented. The prevalence of the symptom in the whole population was 4.8%. Medical literature showed similar prevalence, but considerable variance due to different cut-off values for the symptom.

The prevalence of herpes lips in the *Natrium muriaticum* population was confirmed in Materia Medica Validation, three out 19 patients (16%). Sepia was

Table 11.4 Results for 'Recurrent herpes lips'

Medicine	Herpes lips	No herpes lips	Prevalence	LR	95% CI
Bry	2	7	22%	4.72	1.38–160.9
Carc	3	32	9%	1.82	0.61–5.43
Caust	3	35	8%	1.67	0.56–5.01
Chin	2	11	15%	3.27	0.90–11.79
Kali-c	2	12	14%	3.03	0.83–11.04
Lach	2	28	7%	1.41	0.37–5.42
Lyc	8	64	11%	2.41	1.23–4.71
Merc	3	42	7%	1.41	0.47–4.25
Nat-m	19	110	15%	3.38	2.17–5.28
Sep	8	68	11%	2.28	1.16–4.47
Staph	4	21	16%	3.43	1.38–8.53
Sulph	4	68	6%	1.17	0.45–3.08
Thuj	2	16	11%	2.36	0.63–8.78

not yet validated this way. There was no herpes lips detected in the *Staphisagria* population (n = 18) of Materia Medica Validation.

Because of the significantly positive values the entries of *Bryonia* (bry), *Lycopodium* (lyc), and *Staphisagria* could be added to this rubric. The entries of *Arsenicum album* (ars), *Graphites* (graph) and *Silicea* (sil) are questionable, see later.

Misleading Entries

A clear example of a misleading repertory-entry caused by taking the absolute occurrence of a symptom as guidance is the symptom 'Fear of death' for *Natrium muriaticum*. In the database only one of the 129 patients responding well to *Natrium muriaticum* had fear of death. According to the existing system this would rectify the entry of *Natrium muriaticum* in the rubric 'Fear of death', and this medicine is actually there in the existing repertory. But the LR + of this symptom for this medicine is 0.19. Suppose a case where one estimates the chance that *Natrium muriaticum* will work to be 50%. If this patient appears to have a strong fear of death Bayes' formula will show that the possibility that *Natrium muriaticum* will work decreases to 16%. According to the existing repertory this symptom would increase the likelihood of cure by this medicine.

Comparison LR Data with Repertory

An example based on the original repertory containing six symptoms is presented in Table 11.5. The numbers in the columns below the abbreviations of the medicines indicate the strength of each indication, e.g. the number 3 for the fifth symptom in

Table 11.5 Entries of six repertory rubrics arranged to discover medicines that could be considered in a patient that has these symptoms. Value 1 under the medicine means that the symptom is occasionally seen in patients responding well to this medicine. Values 2 and 3 indicate that the symptom is seen more frequently

Symptom nr.	Symptom	Number of entries
1	RECTUM – DIARRHOEA – anticipation, after	4
2	MIND – FEAR – death, of	145
3	TEETH – GRINDING – sleep, during	43
4	FACE – ERUPTIONS – herpes – Lips – About	44
5	MIND – INJUSTICE, cannot support	10
6	MIND – LOQUACITY	139

	Medicines							
Symptom nr.	Caust.	Ph-ac.	Sep.	Gels.	Calc.	Ign.	Verat	Arg-n.
1	–	2	–	3	–	–	–	2
2	2	2	1	3	3	1	2	2
3	1	–	1	–	1	2	2	–
4	1	1	3	–	–	–	–	–
5	3	–	1	–	1	1	1	–
6	1	–	–	2	1	1	2	–

the first medicine row indicates that sensitivity to injustice is a strong indication for the medicine *Causticum* (caust.).

This system should be handled carefully; all symptom-rubrics mentioned in the repertory were originally constituted separately, as if the symptoms were mutually independent. This is probably not true for a number of medicines. Symptoms should also be chosen carefully, because they could be related. If, say, the patient has rheumatoid arthritis (RA) adding the symptom 'amelioration from motion' gives few extra information because most RA patients have this symptom (Rutten, 2007). Homeopathic doctors use such tables only to concentrate on a limited number of possible medicines to be explored further. The next step consists of pattern recognition. Metaphorically, this could be compared with a weather forecast; a small number of variables like temperature, wind and precipitation can help to make decisions in a far more complex choice about activities for the next day. The three given variables however have different interrelations for each activity.

In comparison with the existing repertory one can now make more reliable estimates about the LRs of various symptoms for a number of medicines. But this outcome has still to be interpreted carefully because numbers vary strongly among medicines. The number of cases responding well to some medicines is given in Table 11.6.

Table 11.7 shows the calculated LRs for these medicines and the assessed symptoms. A blank in this table indicates that there was no patient responding well to this medicine with the symptom present. In bold type are entries that have either not one in their 95% Confidence Interval, or are unlikely to have the LR significantly above 1. In the first case they should be included as an indication for the medicine; in the second case they should be excluded. These entries can be used to estimate

Table 11.6 The number of cases responding well to some medicines

Medicine	N=
Anacardium (anac.)	11
Argentum nitricum (arg-n)	29
Calcarea carbonica (calc.)	64
Causticum (caust)	38
Cocculus (cocc)	10
Conium (con.)	14
Gelsemium (gels.)	10
Ignatia (ign.)	23
Natrium muriaticum (nat-m.)	129
Phosphoricum acidum (ph-ac.)	19
Sepia (sep.)	76
Veratrum album (verat.)	5

Table 11.7 The same symptoms as in Table 11.5, but arranged according to outcome of LR assessment

Symptom nr.	Symptom	Prevalence
1	Diarrhoea after anticipation	4.3%
2	Fear of death	3.9%
3	Grinding teeth during sleep	5.5%
4	Herpes lips	4.8%
5	Sensitive to injustice	9.8%
6	Loquacity	6.7%

	Anac	Verat	Cocc	Gels	Con	Ph-ac	Arg-n	Sep	Ign	Caust	Calc
1	2.12	–	2.33	14.49	1.84	7.61	9.45	0.91	1.01	1.23	1.84
2	13.11	10.42	2.58	2.58	3.71	–	1.78	1.72	2.25	0.67	1.21
3	–	3.65	3.67	–	2.62	1.93	1.90	1.70	2.40	0.48	0.85
4	1.92	–	–	2.11	1.60	1.11	–	2.28	–	1.67	0.33
5	4.69	–	4.12	1.02	–	1.62	–	0.94	2.69	4.17	0.96
6	2.75	9.13	–	–	–	–	0.52	1.60	0.65	1.19	0.46
Total LR	688.2	347.2	90.9	80.5	28.6	26.4	16.6	9.1	9.5	4.0	0.3

the accuracy of the existing repertory by comparing Table 11.7 with Table 11.5. The other entries are rather uncertain, but still based on a more reliable procedure than the original entries in the repertory.

Table 11.7, based on LR, shows similarities to Table 11.5, but also definite differences. The keynotes are correct, each symptom is indeed a (rather) strong indication for the expected medicines. But frequently prescribed medicines, like *Sepia* and *Calcarea*, show the most striking differences with the original repertory. One also sees less frequently prescribed medicines emerging in this outcome, like *Anacardium*, *Cocculus* and *Conium*. Comparing Table 11.5 with Table 11.7, about half of findings do not correspond with the existing repertory. This estimation depend partly on subjective cut-off values for entries in either plain, italics or bold type, but is rather conservative. LR = 1.20 is considered not as contradictory to an existing (plain) entry, although it is not likely to be indicative for a medicine.

LR and Chance

To constitute a repertory with scientific assessment one should regard the influence of chance (Rutten et al., 2008). If one would measure a certain value for LR repeatedly, would it always be the same? Statistics indicate such chances. The LR of the symptom 'Sensitive to injustice' for *Aurum* was LR = 2.28 and for *Nux vomica* LR = 0.68. The chance that the LR for *Aurum* is larger than 1.5 is 85.9%, calculated as exact binomial chance. The chance that the LR for *Nux vomica* > 1.5 is 15.1%.

As mentioned before, semantics could be another problem, especially regarding vague symptoms. 'Sensitive to injustice' can be defined as sensitive to injustice inflicted to others. If handling a broader definition, 'Ailments from anger with indignation' could possibly be included. This symptom is a strong indication for *Staphisagria*. So, before discarding medicines from vague rubrics one should specify such rubrics better. Semantics could be a problem even for a concrete symptom like 'Herpes lips', because there is also the rubric 'Herpes around mouth'. Table 11.8 shows that the 'Herpes lips' is probably no indication for *Arsenicum album*, *Graphites* and *Silicea*. Even considering the probability that LR > 1, one gets p = 0.661 for *Arsenicum album* and p = 0.626 for *Silicea*.

Another reason to be careful with discarding medicines from the repertory is possible interaction between symptoms. In the repertory, as in this assessment, symptoms are considered as independent entities. But it is possible that the combination of two symptoms with low LR is a much stronger indication for a medicine than expected from the LRs separately, see under 'Correlation matrices'. *Repertorisation* could metaphorically be compared with a weather-forecast. One gets variables like temperature, wind and rain as independent values, but the decision about what to do tomorrow depend on a complex weighing of these variables.

Different Cut-Off Values

Considering the medicine *Causticum*, the doctor will ask the patient if he is sensitive to injustice. If he acknowledges the doctor will be satisfied sooner than if he considers a medicine that is not known for sensitivity to injustice. In other words, it

Table 11.8 Some medicines from the repertory-rubric 'Herpes lips' with LR > 1.5

Medicine	Herpes lips	No herpes lips	P-value
Arsenicum album	1	24	0.454
Graphites	0	25	0.154
Silicea	1	26	0.412

Table 11.9 The influence of cut-off values for the intensity of a symptom on LR

Sensitive to injustice		
Higher cut-off value, n = 330	*LR+*	*LR–*
Causticum	4.17	0.669
Lower cut-off value, n = 882		
Causticum	2.45	0.496

makes a difference whether one wants to include or exclude a medicine from the considerations. For exclusion, one will handle a lower cut-off value than for inclusion. In our prospective LR, two cut-off values are considered resulting in three values for the symptom, 0 if the symptom was absent, 1 if the symptom was moderately present and 2 if the symptom was strongly present. For the high cut-off value of 2, the doctors should ask themselves if the symptom was strong enough to begin to think about medicines related to that symptom (including the medicine in the differential diagnosis). For the lower cut-off value of 1, they should be hesitating about the considered medicine if it was absent (excluding). The outcome for the symptom 'Sensitive to injustice' and the medicine *Causticum* is shown in Table 11.9.

As expected, the lower cut-off value includes more patients who are sensitive to injustice. Table 11.9 also shows that the LR + is higher for the higher cut-off value. Intuitively this is easily understood, if the patient is very sensitive to injustice, the medicine *Causticum* is more indicated.

The way to use the absence of expected symptoms is still unclear. Some doctors say that they don't prescribe *Stramonium* if the patient has no fear of the dark. In Materia Medica Validation, however, it appeared that only 42% of the cases that went very well using *Stramonium* (n = 12) had a fear of the dark.

The negative LR (LR-) can be used to exclude a medicine from considerations if the symptom is absent. If the LR- is closer to zero (<1), the medicine is less indicated. It is easy to understand that one hesitates more about *Causticum* if the patient is not in the least sensitive to injustice. The aspect of excluding medicines if the corresponding symptom is absent is as yet not integrated in the repertory. LR assessment enables people to use this aspect, but one should still consider how.

Results of Multivariate Analysis

There were about 1,700 prescriptions with the desired result (GHHOS 2–4), 20 medicines were responsible for 891 of these desired results, see Table 11.10. For these 20 medicines, correlation between two symptoms using correlation matrices and correlation between all symptoms using Principal Component Analysis (PCA) were checked. The discrimination between medicines can be assessed by Canonical Discriminant Analysis (CDA). The number of possible discriminant functions is

Table 11.10 The 20 most (successfully) prescribed medicines

	Medicine	Frequency	Percent
1	Argentum nitricum	24	2.7
2	Arsenicum album	25	2.8
3	Calcarea carbonica	64	7.2
4	Calcarea phosphorica	23	2.6
5	Carcinosinum	35	3.9
6	Causticum	38	4.3
7	Graphites	25	2.8
8	Ignatia	23	2.6
9	Lachesis	30	3.4
10	Lycopodium	72	8.1
11	Mercurius	44	4.9
12	Natrium muriaticum	129	14.5
13	Nux vomica	30	3.4
14	Phosphoricum acidum	19	2.1
15	Phosphorus	55	6.2
16	Pulsatilla	55	6.2
17	Sepia	76	8.5
18	Silicea	27	3.0
19	Staphisagria	25	2.8
20	Sulphur	72	8.1
	Total	891	100.0

limited by the number of symptoms. CDA was performed for 10 medicines that seem related to the investigated symptoms.

Correlation Matrices

Three hundred possible correlations were investigated and hardly any correlations between two symptoms are found. For these medicines, see Table 11.11, the difference between the combined symptoms and the single symptoms was calculated.

The LR of the combined symptoms is calculated by dividing the prevalence of both symptoms together in the medicine population by the prevalence of both symptoms together in the rest-population. The combination of 'Fear of death' and 'Herpes lips' is a stronger indication (LR = 5.39) for *Ignatia* than expected from the LRs calculated for the separate symptoms. If both symptoms were independent their combined LR would be 1.73 * 1.02 = 1.76.

Likewise the combination of 'Diarrhoea from anticipation' and 'Herpes lips' has a higher LR for *Nux vomica*. The combination of 'Sensitivity to injustice' and 'Loquacity' is an unexpected indication for *Phosphor*, as the separate symptoms both do not indicate this medicine. The symptoms 'Herpes lips' and 'Sensitivity to injustice' do not significantly increase the indication for *Phosphoricum acidum*, despite their moderate correlation (correlation = 0.431).

Results of Principal Components Analysis

PCA was used to check if one or more of the symptoms were redundant, because of inter-relatedness with the other symptoms. This is done by making a 'scree-plot', which indicates the amount of information given by each symptom (Fig. 11.2). If the 891 patients in the population responding to 20 medicines could have been described by less symptoms, one would see an 'elbow' in this plot. The elbow indicates that more components give no extra information. This is not the case. Six

Table 11.11 The difference between combined LR for two symptoms after measurement of separate symptoms (LR1 * LR2) and after measurement of the combined symptom (LR(1 + 2))

Medicine	Symptoms	LR symptom	LR1 * LR2	LR(1 + 2)	95% CI LR(1 + 2)
Ignatia	1. Fear death	1. 1.73	1.76	5.39	1.40–20.78
	2. Herpes	2. 1.02			
Nux-v	1. Diarrhoea	1. 1.50	1.17	3.55	1.19–10.58
	2. Herpes	2. 0.78			
Ph-ac	1. Herpes	1. 1.23	1.73	2.19	0.59–8.19
	2. Injustice	2. 1.41			
Phosphor	1. Injustice	1. 0.99	0.92	2.25	1.17–4.33
	2. Loquacity	2. 0.93			

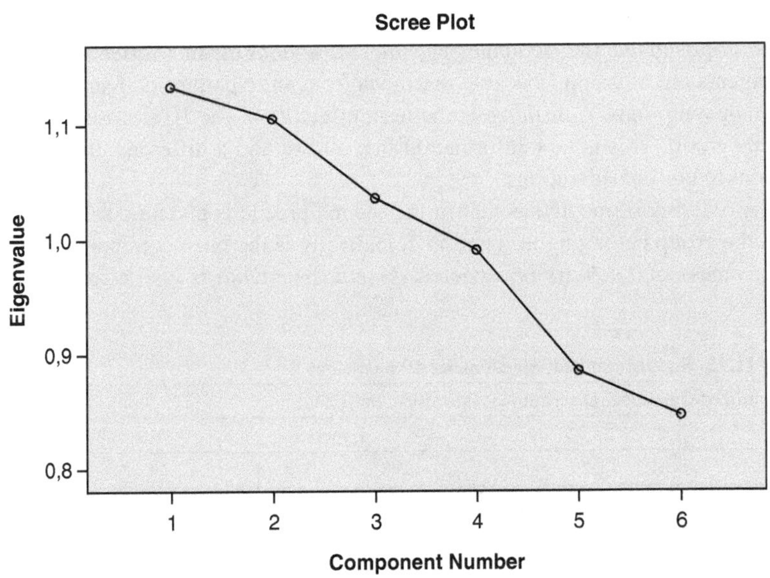

Fig. 11.2 Scree plot indicating the amount of information supplied by each of six symptoms. There is a linear development of information indicating an even spread of information given by each symptom

components are needed to retrieve all necessary information. This indicates that subgroups of patients cannot be differentiated by less symptoms.

Results of Canonical Discriminant Analysis

CDA was performed on the population responding to 10 medicines that seem related to the six assessed symptoms. The results of CDA look much like the results of logistic regression. The first discriminant function of this analysis is:

$$\text{Discriminant1} = 0.867 * \text{diarrhoea} + 0.412 * \text{fear} - 0.069 * \text{grinding} - 0.053 * \text{herpes} - 0.353 * \text{injustice} + 0.227 * \text{loquax}$$

The population consists of ten subgroups, each responding to a different medicine. With six symptoms, one can only discern six subgroups, and that cannot be done by one discriminant function. Dividing the population in six groups, there is a group of patients that have similarities in that respect that they have diarrhoea form anticipation and fear of death; they also tend to loquacity. They do not tend to sensitivity to injustice, herpes lips and grinding teeth at night.

The other discriminant functions can be interpreted in a similar fashion (Table 11.12):

We can also analyse which medicines were effective by each profile and which not (Table 11.13).

The most effective medicine has the highest score. Homeopathic physicians will not be surprised that the medicine *Argentum nitricum* is the most effective medicine for patients out of group 1 because diarrhoea from anticipation and fear of death are the main symptoms. *Causticum* is the least effective of the 10 assessed medicines for this group. This is new information that allows add a differentially diagnostic element to our considerations.

The ordering of medicines that fit the second profile is given in Table 11.14.

In the group belonging to function 2, loquacity is the most common trait, so the effectiveness of *Lachesis* is expected. *Arsenicum album* is also effective in this

Table 11.12 Six discriminant functions for 10 medicines

Standardized Canonical Discriminant Function Coefficients

	Function					
	1	2	3	4	5	6
Diarr	0.867	−0.421	0.291	−0.106	−0.070	0.025
Fear	0.412	0.476	−0.319	0.657	−0.060	0.282
Grind	−0.069	−0.071	−0.037	0.087	0.867	0.499
Herpes	−0.053	0.073	−0.181	−0.402	−0.403	0.807
Injust	−0.353	−0.193	0.663	0.583	−0.169	0.218
Loquax	0.227	0.676	0.614	−0.331	0.137	0.011

Table 11.13 First discriminant function

Discriminant function 1	
Medicine	Score
Argentum nitricum	1.743
Phosphoricum acidum	0.718
Lachesis	0.274
Arsenicum album	0.27
Staphisagria	0.059
Sepia	–0.023
Carcinosinum	–0.074
Natrium muriaticum	–0.303
Mercurius	–0.321
Causticum	–0.377

Table 11.14 Second discriminant function

Discriminant function 2	
Medicine	Score
Lachesis	1.022
Arsenicum album	0.506
Sepia	0.196
Carcinosinum	–0.042
Staphisagria	–0.086
Natrium muriaticum	–0.134
Mercurius	–0.155
Causticum	–0.19
Argentum nitricum	–0.201
Phosphoricum acidum	–0.968

group because of the fear of death. The limitation of CDA in this case is that one can discriminate only six groups because one has only six symptoms.

If one does the same for the other discriminant functions, it is possible to see that *Causticum* works best in function 3 because of the sensitivity to injustice that predominates in this group. The discriminant function 3 shows that *Staphisagria* does not perform well in the group that is most characterized by sensitivity to injustice. See Table 11.15. The CDA does not confirm that the *Stahisagria* patient is more sensitive to injustice than the rest-population, nor does it prove the opposite.

To chose between these 10 medicines combining these three discriminant functions one would prefer *Argentum nitricum* if the patient had diarrhoea from anticipation and was not sensitive to injustice. Translating these data into materia medica, one could represent this medicine like:

Materia Medica *Argentum nitricum*
Positive symptoms: **Diarrhoea from anticipation**, … , … , etc.
Negative symptoms: Sensitive to injustice, … , … , etc.

Table 11.15 Third discriminant function

Discriminant function 3	
Medicine	Score
Causticum	0.716
Lachesis	0.355
Phosphoricum acidum	0.188
Mercurius	0.083
Sepia	0.055
Staphisagria	0.043
Carcinosinum	0.01
Argentum nitricum	−0.023
Natrium muriaticum	−0.277
Arsenicum album	−0.577

Using discriminant analysis one can be more confident about symptoms that indicate the medicine, but one can also use symptoms that contraindicate it, which is an addition to the existing materia medica.

Discussion

The outcome of three different ways of looking at practice experience; consensus procedures by experts, and two different kinds of open cohort studies were shown. In one cohort study, variables (symptoms) were not predefined and analyzed retrospectively; in the other study, the variables were predefined. The most well-known indications for the medicines are correct, but one found considerable differences with the existing repertory, as is expected on theoretical grounds.

Some points for the discussion were already mentioned under the results of the various explored methods. Essential flaws cannot be denied in the proposed method. This is a new research in the homeopathic field, but the methods are not unknown in medicine, especially in diagnostic research. Some modifications were done though to make it work for the daily practice. In conventional medicine, the target of such research is the diagnosis, herein the target was the desired result of treatment. This causes some complications, like handling vagueness and the proper definition of reference (‘gold’) standard ‘the cure’. To calculate LR, different reference groups can be used, one is the population that received the medicine but did not respond to it, the other is ‘all other patients’.

Multivariate analysis was performed on the data from prospective LR research to explore interrelations of symptoms. But another method in multivariate analysis, Canonical Discriminant Analysis (CDA), could offer new possibilities to evaluate clinical data. This research, however, was not designed for CDA. It was only tested regarding its possibilities; other protocols should be used to apply CDA in a proper way.

To improve the repertory with LR research, the software to gather such data should be further improved.

Comparison of Results

The projects were only partly related and the comparisons made here should be considered in that context. Similarities in the outcomes of the different projects were seen with each other and with the existing data, indicating the reproducibility of the homeopathic method. Differences between results of the three projects and the existing data in materia medica and repertory were seen, indicating the structural and stochastic shortcomings of the proposed instruments: absolute instead of relative occurrence of symptoms and limited numbers of observations. Differences between projects due to different methodologies were also seen, especially retrospective and prospective assessment.

To start with the reproducibility of the homeopathic method: at least half of the used data, and especially the data most of participant doctors agree upon, seem correct according to the retrospective and prospective assessments. The Materia Medica Validation showed that homeopathic doctors are rather well able to estimate prevalence of symptoms in their practice populations. The prospective LR assessment and Materia Medica Validation showed about the same prevalence of the demonstrated symptom 'Recurrent herpes lips', but also of the symptoms 'Diarrhoea from anticipation', 'Grinding teeth during sleep' and 'Loquacity'. The prevalence of the symptoms 'Recurrent herpes lips' and 'Grinding teeth during sleep' was confirmed by existing medical literature. The prospective assessment of the symptom 'Sensitivity to injustice' showed how important it is to describe symptoms in the repertory more precisely.

The fact that herpes lips is an indication for the medicines *Natrium muriaticum* is confirmed in all three projects. Its indication for *Natrium muriaticum* and *Sepia* was confirmed in both LR projects. In Materia Medica Validation only a limited number of medicines (not *Sepia*) were validated. Similar results were obtained for diarrhoea from anticipation *vs. Argentum nitricum*, *Gelsemium* and *Phosphoricum acidum*.

The differences between outcomes and the existing materia medica and repertory show that there is considerable room for improvement of the existing instruments. The differences between results and the repertory can be explained by statistical uncertainty caused by small samples and recall or expectation bias in expert experience. However, there are strong evidences that the present data are more correct than the existing entries in the repertory.

The differences between the outcomes of the presented projects can partly be explained by semantics and partly by differences between retrospective and prospective investigation. The repertory has a considerable number of rubrics that are semantically similar, but with different medicines. The rubrics 'Herpes lips' and 'Herpes around mouth' are often used interchangeably, but contain different medicines.

The same goes for 'Sensitive to injustice' and 'Anger with indignation'. In prospective research the investigated symptoms are predefined and checked in every patient, in retrospective research symptoms that are not expected will often not be checked. This could explain why 'Herpes lips' is detected as an unexpected indication for *Staphisagria* by prospective research but not by retrospective research. The same goes for *Anacardium* and 'Fear of death'.

In Table 11.16 retrospective outcome and prospective outcome are compared for "herpes lips". Differences could partly be explained by statistical uncertainty. The prevalence of this symptom is much higher in the prospective research (4.8% against 1.3%). In the retrospective research the symptom could also be divided among two repertory-rubrics, 'Herpes lips' and 'Herpes around mouth', together these symptoms had a prevalence of 2% in the retrospective study. In medical literature large variations in the prevalence of this symptom is found, 3–33% (Shulman et al., 1992; Axell and Liedholm, 1990; Katz et al., 2001; Embil et al., 1975). Most of these differences are explained by different cut-off values. For frequently recurring herpes in the American army and in Israeli soldiers the prevalence is 4% and 5.4% (Shulman et al., 1992; Katz et al., 2001). Differences could also be explained by interpersonal differences and differences between populations. In the prospective research group the prevalence of the symptom varied between 2.4% and 9.8%. Another difference is the varying use of different medicines; in the prospective

Table 11.16 Comparison of results considering the symptom herpes lips between retrospective and prospective LR research

Total number of patients in prospective database = 3,367
In prospective database, 160 patients with herpes lip, prevalence = 4.8%
Total number of patients in retrospective database = 2,148
In retrospective database, 28 herpes lips, 15 herpes around mouth, together 2%

Medicine	LR+ retrospective	95% CI	LR+ prospective	95% CI
Ars	2.02	0.78–4.89	0.84	0.12–5.77
Bry	–		4.72	1.38–16.19
Carc	–		1.82	0.61–5.43
Caust	0.49	0.07–3.58	1.67	0.56–5.01
Chin	–		3.27	0.90–11.79
Kali-c	–		3.03	0.83–11.04
Lach	–		1.41	0.37–5.42
Lyc	0.24	0.03–1.81	2.41	1.23–4.71
Merc	1.08	0.21–10.99	1.41	0.47–4.25
Nat-m	2.17	0.83–5.59	3.38	2.17–5.28
Nux-v	1.49	0.42–5.26	0.70	0.10–4.83
Sep	2.49	1.02–6.07	2.28	1.16–4.47
Sil	1.34	0.41–4.38	0.78	0.11–5.36
Staph	2.67	0.36–19.94	3.43	1.38–8.53
Sulph	0.59	0.14–2.47	1.17	0.45–3.08
Thuj	1.59	0.22–11.44	2.36	0.63–8.78

research group the frequencies of use of medicines varied considerably. The most well-known medicines for herpes lips, however, *Natrium muriaticum* and *Sepia* show similar results.

The retrospective data of practice registration and materia medica validation overlap just a little. The practice registration is a systematic recording of all *repertorisations* by one doctor, the materia medica validation is based on retrieving data in about 14 practices by hand search of patient cards. In the first case repertory language is reproduced, in the second the own wordings of the doctors. This leads to the problem of vagueness of clinical symptoms.

Vagueness

According to classical logic a term is vague if it has borderline cases (http://plato.stanford.edu/entries/vagueness). This is described by the 'Sorites paradox' (http://plato.stanford.edu/entries/sorites-paradox). How few hairs should someone have to call him bald? Vagueness is inherent to medical practice. The clinical notion of vagueness comprises also *generality* and *ambiguity*. The Sorites vagueness is quantitative, whereas generality and ambiguity are qualitative.

The concept of generality can be illustrated in the repertory by some common rubrics, such as 'Loquacity'. Several sub-rubrics like 'Loquacity, during which answers no questions', or 'Loquacity, changing quickly from one subject to another', or 'Loquacity, makes speeches' are also present in the repertory. It is important to consider that 'making speeches' is not some sort of 'loquacity'. Thus, 'loquacity' can be considered an ambiguous notion (Rutten et al., 2003).

Vagueness is predominant in homeopathy because choosing a medicine is quite complex. It depends on the co-occurrence of a number of symptoms, which can change the importance of a first observed symptom. If *Lachesis* is prescribed, for example, it is possible that one will be induced to consider the patient as loquacious. However, if another medicine would be prescribed to the same patient, maybe the symptom "loquacity" would not be considered in the whole observation. Thus, the margin for error based on quantitative vagueness can easily lead to different interpretations, caused by expectation bias. More details about quantitative and qualitative vagueness is found in Rutten et al. (2003).

Gold Standard

The LR + of a symptom regarding a homeopathic medicine indicates the increase of likelihood of a curative action of that medicine. This 'test' is not meant to diagnose an illness but a curative potential of a medicine. So 'cure' caused by the investigated medicine should be the gold standard (GS) (Rutten et al., 2004a). It is impossible to define 'cure' in a clear and unambiguous way, therefore one must find an operational definition for cure.

In homeopathy, 'curative action' ('cure') is not the disappearance of symptoms. Although there is a general consensus about the meaning of 'cure' after homeopathic treatment (Kent, 1900; Swayne, 1998), several scales describing the outcome of treatment were done; one of them was developed in Glasgow (GHHOS). On the other hand, for the prospective LR research, a Dutch specification of this scale, constituted at a consensus meeting of 80 Dutch homeopathic physicians in 1996, was used (Stolper et al., 1998).

The causal relation between cure and medicine is also assessed in this process.

In comparison to the Glasgow version, in the Dutch version two things are specified (Rutten et al., 2004a):

1. The effect is related to natural course of illness, premorbid course and gravity of illness (acute situation).
2. Positive or negative external influences on the course of illness (chronic situation).

The estimation of causality relies strongly on clinical judgment. One uses a subjective threshold value to distinguish if the effect is caused by a medicine, or by other factors, or maybe not cured at all. Other problems, like ignorance about the difference of effects of different potencies are not considered here.

Imperfections of the gold standard have greater influence, as the target population is smaller in relation to the remainder of the population. This is the case in homeopathy because one medicine 'fits' only a small part of the population. The largest population responding to one medicine was the *Natrium muriaticum* population (n = 129) in a total population of 3,367. Suppose to observe a population of 1,010 patients after treatment, and 10 (1%) are in reality cured by a certain medicine. Some patients seem cured according to our gold standard (GHHOS scale), but are in fact not cured by the medicine but by other causes. In this case our gold standard shows us a false positive (FP) result. On the other hand, some cures are not detected by the GHHOS, suppose the GHHOS detects only 80% True Positives (TP).

Because the population that is in reality not cured by this medicine (rest-population) is so much greater the cured population, mistakes in the rest-population have far more influence than mistakes in the cured population. In the observed population only 80% of the cured patients are recognized and only 1% of the patients not cured by the medicine is falsely recognized as cured (FP). In spite of the far better filter for false positives, this GS leads to 10 false positives and eight true positives. Similar calculations can be made for the remainder of the observed population, where two patients will show up who are in fact cured by the medicine, false negatives (FN). Calculating the influence of FP and FN arising from gold standard on LR would be possible if these factors were known.

Generally, one doesn't know the amount of FPs and FNs arising from our gold standard, but one should realize that it can lower the LR + values considerably if the remainder of the population is much larger than the population responding to a particular medicine.

Influence of Vagueness and Gold Standard on LR

The showed results are significantly influenced by vagueness and gold standard. One should realize, however, that the existing situation is much worse. About half of the entries in frequently used rubrics of the repertories is different from our results. Even if the presented results are partly incorrect, they are not as incorrect as the existing repertory rubrics, since these rubrics were included from less rigorous observations.

An important outcome of these theoretical considerations is that doctors participating in this kind of research should know of these problems. They must realize that their bias weakens their own instruments. If they try to prove their own opinion, e.g. by indicating all patients as 'sensitive to injustice' once they consider *Causticum,* they will flaw results. This could work two ways. If *Causticum* really works, the importance of the symptom for *Causticum* is over-estimated. If not *Causticum* but another medicine works, this medicine increases the number of 'sensitive to injustice' patients in the rest-population. In the first case the LR of this symptom is over-rated; in the second case the LR is under-rated. It can be shown that false positive results lead to under-rating of LR.

It is hard to say which influence bias has on results; ways to discover this must still be found. Based on the theoretical considerations bias could work both ways, over-rating as well as under-rating. But this problem is particularly relevant if one wants to assess the absolute value of several symptoms regarding one patient. Probably comparing different medicines out of one symptom-rubric is less sensitive to bias because the symptom is assessed the same way for all patients.

Confirmation Bias

The problem mentioned above that doctors tend to see 'Sensitivity to injustice' sooner if they have other reasons to consider *Causticum* is called confirmation bias. This was actually the case in the prospective LR assessment. After a few months there were no patients responding well to *Causticum* that were not sensitive to injustice, so LR + was infinite. Later on, patients who were not sensitive to injustice nevertheless appeared to respond well to *Causticum*, so LR + lowered to 4.17 after 33 months.

Vagueness in symptoms like 'Sensitivity to injustice' and 'Loquacity' could lead to confirmation bias. There could also be a difference in this respect between retrospective and prospective research. One cannot compare our data about 'Sensitivity to injustice' because of differences in interpretation, but the results for 'Loquacity' are the same. In the retrospective LR assessment, the relation between loquacity and *Lachesis* was found to be LR = 5.37 (95% CI 2.5–11.8) and in the prospective assessment LR = 5.20 (95% CI 3.08–8.76). In the prospective assessment, data were gathered on regular intervals and gave feedback to the investigating doctors. This could influence results.

Recall Bias

Recall bias is the ignoring of a symptom that is actually there. Possibly the patient has forgotten it, but doctors cannot verify every symptom. This is more likely to happen in retrospective research because doctors don't check symptoms that are far away from their prior considerations. In the prospective research some unexpected results were seen, like LR = 12.11 (6.20–23.65) for 'Fear of death' and *Anacardium* and LR = 4.19 (1.69–10.42) for 'Herpes lips' and *Staphisagria*. These symptoms are not known to be relevant for these medicines and doctors will not check these symptoms when they consider these medicines.

Recall bias in retrospective investigation becomes larger as doctors record less symptoms per patient in their computer program. In the retrospective LR research an average of 6.5 symptoms per patient (range 0–19) were recorded.

Which Control Group?

In the LR project LR + is calculated as the relation between the occurrence of the symptom in the target population (= cured by medicine x) and the occurrence in the rest population (= people not cured by medicine x, people cured by other medicines and people that could have been cured by medicine x but who did not get this medicine). Some people say that one should compare people where a certain medicine worked with people where this medicine was prescribed and did not work. These different views reflect a difference between homeopathy and conventional medicine. In homeopathy, the possibilities for several different medicines are considered for the same diagnosis. In conventional medicine, one medicine works for that indication, or not.

Bayes' formula is available in several variants, one of them is:

$$P(H_i \backslash E) = \frac{P(H \cap E)}{P(E)} = \frac{P(E \backslash H_i) * P(H_i)}{\sum_{allhypotheses} P(E \backslash H_i) * P(H_i)}$$

Where H_i = hypothesis that medicine x works
E = evidence, in this case the symptom

According to Bayes formula the hypothesis that one medicine works if the symptom is present should be compared with the hypotheses concerning all other medicines. This is consistent with the homeopathic algorithm. But there is inaccuracy because prescriptions are not infallible.

Software Requirements; Incidence or Prevalence

As doctors want to gather a large amount of data during consultations, the role of software is vital. Recording of symptoms, prescribed medicines and results should take as few time as possible and cause no mistakes. Database programs are the best means for this purpose, especially programs that are already used for administrative purposes during consultation. These programs already comprise the personal data of the patient and can be at hand during consultation.

Several database-programs for administration of homeopathic consultations have one important limitation: they record every prescription of a medicine as a separate record linked to each consultation. This way the incidence of the prescription of each medicine is obtained, not the prevalence. If the patient comes back many times and receives always the same prescription, this medicine will be over-rated in the database. For proper administration of results computer-programs must be able to produce the prevalence of each medicine-prescription. It is also essential that doctors have an overview of the medicines they prescribed the patient sitting before them. They should be able to correct results connected to each medicine if that medicine seemed to work at first but appears to be ineffective after all, or if the effect builds up.

Another important point is the proper default value for each variable. The default value for result should be a blank. If the default were a zero one would get zero results for all patients lost to follow-up. This influences LR values. As the method is not to be proved, an intention to treat analysis is not been performed. LR calculations should be based on observable values. For recording symptoms in prospective research, however, the default value can be zero.

Possibilities for Discriminant Analysis

The multivariate analysis of this database had two goals:

1. Detect interrelations between symptoms
2. Detect patterns of symptoms in our population

For the first goal, correlation matrices were made and PCA was made. Any correlation between symptoms for the 20 most frequently used medicines was hardly found. PCA gave no indication that the number of symptoms could be reduced for these 20 medicines.

The LR project was designed to compare symptoms, not medicines. The LR results offer new possibilities. Homeopathic physicians use present as well as absent symptoms, but their instruments (repertory and materia medica) refer only to symptoms that can be expected and only if the symptom is present. LR gives indications about symptoms not to expect when a certain medicine is considered e.g. herpes lips considering *Graphites* ($p = 0.154$). Negative LR also gives the possibility to estimate the meaning of the absence of an expected symptom.

LR assessment is a long-term project. One needs to assess more than 500 symptoms to get a rather complete evaluation of the most frequently used homeopathic symptoms. This will improve the method step-by step, but takes a considerable amount of time. One of the advantages of LR assessment is that any medicine can turn up to be useful if a particular symptom is present.

Discriminant analysis is suited to maximize differences between different groups (medicine populations) by a combination of present and absent attributes. Discriminant analysis allows evaluate a number of medicines at the same time. If one evaluates the 20 most prescribed medicines, half of our cases will be improved. CDA will give information about which symptoms to expect and which symptoms not to expect when consider a certain medicine. It gives, however, no information about other medicines and one cannot differentiate these 20 medicines from other medicines. CDA does not give the information of negative LRs; the meaning of the absence of an expected symptom is not calculated.

CDA was performed on the populations out of the prospective LR investigation responding to 10 medicines and evaluated six symptoms. Each symptom is considered to be a keynote for different medicines. This is not ideal for this purpose, one should have had more symptoms than medicines. But it is still shown that the *Argentum nitricum* patient is unlikely to be sensitive to injustice and that the *Causticum* patient is unlikely to have fear of death.

Discriminant analysis is an interesting option to improve homeopathic prescribing. CDA should be considered for evaluation of cured cases considering the 20 most prescribed medicines. Each medicine has a number of 'keynotes', symptoms that occur rather frequently in the population that responds well to that medicine and are the most indicative for that medicine. CDA would be effective if one assesses a number of symptoms for each medicine. If five keynote-symptoms is chosen for each of 20 medicines, this would constitute a questionnaire of 100 symptoms (maximum). If that questionnaire is given to a sufficient number of patients that responded well to the investigated medicines, a reliable differential diagnosis of the most successful medicines could be gotten.

Future Developments

Until now, enormous amounts of practice experience are lost, despite the new possibilities of computers. Repertory and materia medica can be improved if statistical knowledge and instruments are applied. Here, three possibilities are depicted.

With Discriminant Analysis the 20 medicines that bring on half of successes could be assessed. It should be possible to interview an adequate number of patients that responded very well to these 20 medicines within a few years. Then, the symptoms indicating these medicines can be accurately known, but symptoms that contraindicate these medicines are also added.

With prospective LR research, one will be able to discard superfluous and misleading entries of the repertory and have more accurate comparisons between medicines indicated by the same symptom. Unexpected medicines that are indicated by the assessed symptoms will be also seen. A new possibility of LR, to be explored further, is the use of negative LR to estimate chances that a medicine will work if an expected symptom is absent. Prospective LR research is best suited for keynote symptom with a prevalence of 5–10%. Some 600 of such symptoms are estimated. With 50 research groups, it will take eight years to update the repertory to a large extent.

Every homeopathic doctor should record what he or she is doing. With retrospective analysis of thousands of databases one confirms (or not) the outcome of the other projects mentioned here and get reliable data of peculiar symptoms. One gets indications of LR values of all rubrics in the repertory. And last but not least, one can retrieve the best cases to reach consensus about the complete picture of each medicine.

Acknowledgements The retrospective Materia Medica Validation was supported by the SHO (Dutch homeopathic post-grade training for doctors). The prospective LR research was sponsored by KVHN (royal Dutch patient's organization for homeopathy), the Louise van Eighteen foundation, SFWOH (foundation for homeopathic research) and VHAN (Dutch homeopathic doctor's association). The doctors participating in the prospective LR-research were Rob Bethels, Hefty Butler, Paul Fruitier, Gerard Jansen, Jean Pierre Jansen, Stan Jesmiatka, Christien Klein, Roland Lugten, René van der Reijden, Lex Rutten, Erik Stolper, Janny Verhey and Mechtild Wijdeveld. Fragments and derivatives of the whole idea are published in: Homeopathy 2002, 91:230–238; Homeopathy 2003, 92: 213–216; Homeopathy 2003, 92: 182–186; Homeopathy 2004, 93: 120–124; Homeopathy 2004, 93: 190–192; Homeopathy 2004, 93: 78–83; Homeopathy 2006, 95: 88–93; Homeopathy 2007, 96: 120–124; Homeopathy 2008, 97: 16–21.

References

Axell T, Liedholm R. Occurrence of recurrent herpes labialis in an adult Swedish population. Acta Odontol Scand; 1990, 48: 119–123.

Bayes T. An essay toward solving a problem in the doctrine of chances. Philos Trans R Soc Lond; 1763, 53: 370–418.

Embil JA, Stephens RG, Manuel FR. Prevalence of recurrent herpes labialis and aphthous ulcers among young adults on six continents. Can Med Assoc J; 1975, 113: 527–630.

Katz J, Chaushu G, Peretz B. Recurrent oral ulcerations associated with recurrent herpes labialis – two distinct entities?. Community Dent Oral Epidemiol; 2001, 29: 260–263.

Kent JT. XXXV. Prognosis after observing the action of the remedy. Lectures on homeopathic philosophy. Wellingborough: Thorsons Publishers; 1900.

Rutten ALB. Bayesian homeopathy: talking normal again. Homeopathy; 2007, 96: 120–124.

Rutten ALB, Stolper CF, Lugten RF, Barthels RJ. Assessing likelihood ratio of clinical symptoms: handling vagueness. Homeopathy; 2003, 92: 182–186.

Rutten ALB, Stolper CF, Lugten RF, Barthels RJ. 'Cure' as the gold standard for likelihood ratio assessment: theoretical considerations. Homeopathy; 2004a, 93: 78–83.

Rutten ALB, Stolper CF, Lugten RFG, Barthels RWJM. Repertory and likelihood ratio: time for structural changes. Homeopathy; 2004b, 93: 120–124.

Rutten ALB, Stolper CF, Lugten RFG, Barthels RWJM. A Bayesian perspective on the reliability of homeopathic repertories. Homeopathy; 2006, 95: 88–93.

Rutten ALB, Stolper CF, Lugten RF, Barthels RJ. New repertory, new considerations. Homeopathy; 2008, 97: 16–21.

Shulman JD, Carpenter WM, Lewis DL. The prevalence of recurrent herpes labialis during an army hot weather exercise. J Public Health Dent; 1992, 52: 198–203.

Stolper CF, Kipp RP, Lugten RFG. A proposed consensus on evaluating a case: results of the VHAN-conference. Homeopathic Links; 1998, 11: 51–53.

Stolper CF, Rutten ALB, Lugten RFG, Barthels RJWM. Improving homeopathic prescribing by applying epidemiological techniques: the role of likelihood ratio. Homeopathy; 2002, 91: 230–238.

Swayne J. The Response to the Prescription. Homeopathic Method. New York: Churchill Livingstone; 1998.

Swen JJ, Huizinga TW, Gelderblom H, de Vries EGE, Assendelft WJJ et al.. Translating pharmacogenomics: Challenges on the road to the clinic. PLoS Med; 2007, 4(8): e209. doi:10.1371/journal. pmed.0040209.

Part V
Veterinary Research and Practice

Chapter 12
Homeopathic Veterinary Clinical Protocol

Luiz Figueira Pinto

General

The homeopathic knowledge has traditionally been run by autonomous professionals who transmit their experiences and become the point of reference of practices to be followed, supported by a literature built in the same way (Thompson, 2004). This knowledge has been taken from an old socio-cultural and temporal context and introduced in the nowadays technical-scientific context, which involves intellectual discrepancy and epistemological consequences (Pfuetzenreiter, 2002), contributing to and an ardent sectarianism toward homeopathic knowledge. In the veterinary practice, this pedagogic context gets worst when some professionals add an anthropomorphic view when doing clinical interpretations (Pinto, 1997, 1998a).

The pedagogic procedure to educate and train professionals of medical field requires logical and objective criteria for a proper development of skills and of medical praxis. The maintenance of homeopathic teaching in academic environment includes generation of knowledge and an interdisciplinary work.

The first obstacle that appears in the creation of the academic homeopathy is the scientific communication. The peculiar technical jargons of homeopathy, the heritage from an era of vitalistic thought, do not go by in the collective imaginary today, even if the explanations are good. Another obstacle is the subjectivity of the classical homeopathic rationality, limited by a temporal knowledge and built by a propaedeutic that values the report of the patient and prioritizes, in a rigid way, the so called mental symptoms rather than the physical symptoms. The simplicity of clinical practice contrasts to the whole theoretical principles that, in general, use the comparison of the patient's subjective clinical report, to choose the medicine to be prescribed. And it takes no counts of the information contained in the medical field, ignoring the mechanism of action and the physiopathological process involved (Pinto and Almeida, 2002a).

Universidade Federal Rural do Rio de Janeiro, Rio de Janeiro, Brazil. luizfigueira@ufrrj.br

Presented at the XXth Symposium of GIRI, São Paulo, 2006

L.V. Bonamin (ed.) *Signals and Images*,

With this concern in mind and immersed in the purpose of helping in the resolution of these issues, in 1992 the Homeopathic Unit in the Medical Clinic of the Veterinary Hospital at *Universidade Federal Rural do Rio de Janeiro* (UFRRJ) was created, offering practical activities to the students enrolled in a *lato sensu* post-graduate course about homeopathy at the *Instituto Hahnemanniano do Brasil* (IHB). In 1996, with the recognition of homeopathy as a veterinary medical specialty, the subject "Veterinary Homeopathy" was inserted in the graduation *curriculum*, training professionals for the practice of homeopathic therapy in different clinical and surgical situations for pets, wild and production animals.

The educational planning of these activities is supported by different research lines, providing the generation of knowledge established from epistemological criterion and looking for a rational policy for the understanding of the homeopathic phenomenon (Pinto and Almeida, 2002a).

A homeopathic clinical protocol was developed from the concepts described by Pinto (2001) in order to allow the patient care in different clinical situations, using comprehensible clinical criteria for the scholars and the professionals, and that demonstrates applicability and standardization of homeopathic practice, which contribute to its validation. The consequence of this approach is the interaction of homeopathy as a therapeutic modality with other veterinarian medical specialties and its recognition as a field of research.

The Homeopathic Clinical Protocol

The clinical protocol used at the Homeopathic Unit in the Medical Clinic of UFRRJ Veterinary Hospital is structured in a pragmatic medical rationality, whose propaedeutic approach allows the collection of signs and symptoms according to clinical criteria and established benchmarks. One of the purposes of this standardization is to obtain data in an objective way so that the diagnostic conclusions and the therapeutic conduct are identical among the members of the medical team, and can be applied in different animal species (Pinto and Almeida, 2002a, b; Pinto, 1998b, 2000, 2001; Almeida and Pinto, 2002; Castilhos et al., 2002, 2003). As a complement, the therapeutic decisions can be justified within this rationality, the prognosis may be issued and the clinic evolution can be evaluated and compared with the previous diagnostic conclusions.

The Clinical Form

The execution of the protocol requires the registration of all assessed data relevant to the irregularity in patient health in an appropriate clinical form. In addition to all the components of a classic propaedeutic form, as identification and clinical history, in this homeopathic clinical form there are the diagnostic findings to be issued, describing the clinical condition of the patient at that moment. These data will be used as a benchmark to establish the therapeutic conduct to be followed.

The mental, general or particular characteristics, shown by symptoms, can appear on the item form that is more relevant to you. This clinical history is based on anamneses, either free or directed, which seeks to identify the physiopathological mechanism responsible for the health irregularities.

The whole clinical history is complete with the clinical examination, whose objective data lead to draw diagnostic conclusions that will guide the homeopathic prescription. At the end of this procedure, the conduct adopted is justified according to organic compatibilities that must be consistent to the comparison between clinical picture (of patient or illness) and repertorial and pathogenetic images (Pinto and Almeida, 2002a; Pinto, 2001).

This clinical homeopathic protocol provides the following diagnoses:

a) Clinical diagnosis
b) Dynamic-clinical diagnosis
c) Biopathographic diagnosis
d) Biotypologic diagnosis
e) Temperament diagnosis
f) Diathesic diagnosis
g) Midical diagnosis

The Clinical Diagnosis

The clinical diagnosis consists of all the elements that are involved in health disturbance. This can be set up as an established nosologic entity, with identified etiopathogenesis, its classical medical denomination and its clinical picture, represented by signs and symptoms. These clinical signs and symptoms represent the "minimum syndrome of maximum value", the symptomatic totality, which, in addition to other clinical diagnoses, represents the image of patient and illness. These events will be used as a benchmark in the process of healing.

With the establishment of the symptomatic totality, the Hahnemann's disease classification is, then, made. In cases of extrinsic causality disorders, named as non-dynamic, such as the majority of surgical diseases, the overloads and poisoning, the conduct is fast and can abstract from other diagnoses. Drugs called circumstantial, which have in its toxicology the clinical correspondence by similarity, accomplish the prescription in such cases. In cases it can fit in an elective surgery, the prescription of *Arnica montana* is always indicated (Pinto, 1998b). In non-elective surgery, with certain surgical diseases involving handling and tissue or organs loss, one can include other drugs, such as *Calendula officinalis*, because of the supurative wound nature; and *Silicia*, because of the presence of drains (Pinto, 2000).

The intrinsic causality disorders, named as dynamic, including acute and chronic diseases, require other diagnoses to complement the clinical picture. In acute cases, either epizootic or enzootic, the clinical image will be composed by the disease signs and symptoms, not patient signals and symptoms. The selected medication, which is commonly called "epidemic genius", must show biotypologic and diathesical compatibility with the disease physiopathology (diathesis refers to the predisposition of contracting different diseases, from a similar nature and with hereditary character). For example, the use of *Phosphorus* is indicated to the treat dog *Erlichia infections,* because it is compatible to the phosphoric biotype and the syphilitic diathesis (Pinto, 2001).

The chronic and diathesical diseases are the most frequent in the clinical routine and require the study of the present disease history, the previous pathological history, the biopathographic, biotypologic and temperamental diagnoses (Pinto and Almeida, 2002a, b; Almeida and Pinto, 2002; Castilhos et al., 2002, 2003).

The Dynamical Clinical Diagnosis

This diagnosis reflects the evolution of the morbid process, which can be classified as sensorial, functional and lesional. The registration is done through the most advanced level, even considering that the previous phase is commonly present. The diagnostic conclusion of this clinical evolution must occur in the first consultation, and not in the following or in a return, as the forecasts of Kent suggests (Pinto, 2006). The current conventional medicine uses a series of additional tests that limit the clinical evaluation, mainly to lesional stage, by not treating the initial stage of the morbid process.

Sensory manifestations always occur and can provide important benchmarks for the homeopathic medical indication. The clinical manifestation in the sensorial plan is subjective and with less prosperity in veterinary medicine, directing to anthropomorphic interpretations, which may lead to error. Only symptoms that can be truly confirmed by the examiner and his team as, for example, manifestations of painful feeling, indisposition, irritability, aggression, possessiveness, joy, weakness, itching, and others, can be considered and registered at the homeopathic form.

The functional morbid process is common and in much of the cases is the main complaint of the clinical history, such as fever, diarrhea, vomiting, cough, secretion, and other organic disorders. These clinical manifestations are objective and can indicate the dominance of a working system in which the animal organism is exposed (Pinto, 2007). The occurrence of certain clinical signs such as diarrhea denotes an irregularity in the state of health, which results in greater difficulty of maintaining certain internal constants, in which the body establishes a functional hierarchy, with mobilization and prioritization of certain functions. This is a pathophysiological moment, in which there is a restricted organic balance, with reduced thresholds of adaptive responses, and a greater metabolic spent. All the individual attention is faced to the morbid process, at first with great energy mobility, featuring a wasting stage where inflammation, fever, perspiration and agitation can occur. This is followed by a functional exhaustion, in which the clinical picture moves to an asthenia stage, with less metabolic spent and loss of life quality.

This pathophysiological context, where the organism reveals one of its sick part, with functional hierarchies directing the metabolic activities with the participation of the whole organism, denotes the existence of an organism composed of parts which interact, where each functional unit works for all others, and all of them work according to the affected part, in an hologram way (Pinto, 2007).

The medical interference must respect this pathophysiological moment and the selection of medication must be oriented to local clinical signal manifestations and their corresponding feeling, as advocated in the Boenninghausen technique (Pinto, 2006). In these aspects, one should look for a product that matches the asthenia stage. For example, *Lycopodium clavatum* is a product that fits into various wasting manifestations, revealing loss of liver and gastrointestinal activity, dry and wrinkled skin and physical and mental weakness. *Aconitum napellus* demonstrates apoplectic asthenia features, reveling hyperemia, inflammation, spasms and agitation (Vijnovsky, 1980).

The lesional manifestations occur with great frequency in veterinary routine. They reveal the commitment degree of the organ or tissue involved, and can be classified as light, serious or incurable lesion. At this stage of clinical development, an inappropriate approach may encourage the appearance of clinic aggravations by emunctories blocking. A surgical procedure may be required, as occur in case of neoplasia.

The light lesional stage commonly arises in superficial organs to the level of skin or mucous membranes, in the form of eczemas, skin ulcerations, alopecia, blebs, and others. The recovery of this stage can simply occur with the use of circumstantial medicines, drainers and organotherapics for internal use, associated or not to a topic treatment (Pinto, 2006). The emunctories functions should be improved, in order to ameliorate the drainage function. A drainer medicine commonly uses psoric mechanisms and, when this is the only diathesis involved in the morbid process, the healing process takes place quickly, without aggravations. The cases of non-clinical improvement may reveal the inappropriate use of diathesis incompatibility medication.

In severe lesional stages, the hospitalization for the use of other therapeutic modalities may be necessary and the prognosis is commonly reserved. The organic conditions for response to homeopathic medical stimulus earn some limitation. The main clinical situations in this stage can be, for example, profuse bleeding, gastrointestinal ulcers, internal inflammatory processes, such as encephalitis, hepatitis, nephrites, among others. A patient who presents previous castration history can develop to this stage because of the lack of drainage via. An example is the urological feline syndrome, which occurs more frequently in the castrated cat (Taravenic and Pinto, 1998).

Incurable patients are those who do not respond to homeopathic medication because of the malfunction of endocrine glands, liver or kidneys, indicating severe emunctories blocking. This stage imposes therapeutic limits and requires the use of symptomatic and maintenance medicines to ensure a minimum life quality. Some pathological processes, such as cancer, are present in this stage. In a neoplastic process, the tumor staging must show the therapeutic modality to be followed, and if there is surgery indication the tumor extraction can save the patient, but the clinical recovery depends on the treatment of the diathesical chronic illness.

The Biopathographic Diagnosis

The predisposing factors of the current disorder are regarded as determinant elements of the biopathography by revealing the animal sensitivity to certain agents and tendency to illness.

Currently, the excess of antigenic vaccine stimulus constitute relevant morbid factor for determining immunogenic overload, and some diathesic diseases may be associated with this element, but studies in veterinary medicine clarifying this mechanism are still missing.

The inadequate management, as excessive bathing, use of disinfectants, inadequate diet, lack of physical activity, sexual abstinence, among others, is relevant factor in the unleashing of organic disorders in animals. Its identification may require the adequacy or the suspension of certain practices, so that the process of healing can occur and the organic balance may be established. The biopatography refers to disorders occurring over the life of the patient, on the previous pathological history, including the predecessors' diseases and those recorded in the family history.

The current diathesic illness must be analyzed together with the diseases that the patient presented throughout his life, to conclude on present diatheses or whether they differ from the diatheses manifested by the family predecessors. The manifestation of common diathesic reactivity on the family lineage denotes adaptation compatibility and is easily therapeutically driving, while the manifestation of an unusual diathesis to the family lineage may set a diathesic incompatibility, with consequent lack of specific organic skills and greater seriousness of the case. This procedure is essential to the correct prescription, because it considers that the diatheses are inherited with a structural adaptive capacity, which makes treatment easier. The existence of a familiar diathesic incompatibility requires an initial treatment directed to the diathesis (Carillo Júnior, 2000).

In veterinary medicine the knowledge of some breed or strain predisposition can facilitate this record, and the incurability character of certain genetic diseases should be considered. Just one unidentified biopathographic factor may result in therapy failure and the incorrect interpretation of the whole healing process.

The Biotypologic Diagnosis

The animal morphological characteristic is genetically inherited and determines physiological and behavioral trends. This inherited model suffers some structural adaptive changes throughout life according to its previous pathological history.

This knowledge is of great value because the morphological and functional animal characteristics are reflected in its medical specificity and, therefore, can direct the therapeutic conduct.

The shape of the body and its physiological and behavioral characteristics constitute the constitutional aspects of organisms (Carillo Júnior, 1997). In veterinary medicine the morphology study has been done by anatomy and zootechny, unrelated to medical clinic. In orthodox clinical practice there is no relationship between biotypology and therapy, just the body index or animal weight are used as benchmarks of health. However, in homeopathic therapy this knowledge is extremely useful.

The biotypology concept refers to that presented by Henri Bernard, in 1947, and it is related to the mineral metabolism that predominates in the individual body. These mineral constituents actively participate on both the structure and cell physiology and are related to certain morphological and functional characteristics. The elements sulphur, calcium, phosphorus and fluorine are identified in this case. Consequently, human beings are related to three basic biotypes: sulphuric, phosphoric and carbonic. The three biotypes can receive the anatomic scars, which determine what is called "fluorism". In practice, it is observed that commonly individuals show a biotypologic mixture depending on the family legacy (Carillo Júnior, 1997).

The animal clinic condition in a specific time will depend on its biotypology, which can be correlated to certain medicines, the so-called constitutional medicines. The constitutional medicines are formed by the reaction of a mineral acid with a base, where the acids are related to the biotype, such as sulphuric, carbonic and phosphoric acids and the bases correspond to certain specific clinical states. Although there are practically no systematic studies about the efficacy of this medical approach, this reasoning has shown to be very useful in the UFRRJ Hospital routine (Pinto, 1998c).

It is accepted that morphological and behavioral characteristics are resultant from animal functional and structural adaptations to certain environmental conditions. It could be considered, therefore, that each biotype is better adapted to a particular ecological environment. This situation could occur among wild animals, which always seek the most favorable environmental condition for them, including those that promote migration in search of that most suitable environmental condition. The human interference in the cross processes of domestic and production animals could have provided the development of different biotypes within the same species and the occurrence of a biotypologic mixture or biotypologic races.

Using this reason, dogs would be animals that often show mixed biotypes, with a sulphuric, carbonic or phosphoric constitutional basis, because of the great diversity of races carried by man and their cosmopolitan habits. The fluorism present in dogs would be similar to that seen in humans. As examples of biotypes in dogs there would be the high degree of phosphorism from Whippet race, and the high-degree of carbonism in Napolitan Mastiff and St. Bernard races. As sulphuric animals, a good example would be the Beagle race.

The equines seem to have been originated from a constitutional sulphuric basis. And, currently, reveal under morphological, physiological and pathological aspects, greater or lesser degree of phosphorism, depending on the race. For example, the animals from the Arabian race and Mangalarga are well fitted in sulphuric biotype, while the animals from the English Pure Blood race show a certain degree of phosphorism. The sulphuric constitution in these animals provides a good muscle mass distribution and a proportional body, with adequate capacity to adapt to environmental adversities. The tendency to poisoning and consequent congestion determines the need for physical exercises, while the phosphorism is manifested by a higher body growth, an increased thyroid activity and fast wears of the suprarenal. These animals are more agitated and have quick responses.

Cattle are animals that show mixed biotypes, which are distinguishable by the Indian or European origins. The Indian races show predominantly catabolic-metabolic features, with a certain degree of phosphorism under a sulphuric base. An example is those animals from the Nellore race, which are thin and with agitated behavior. Animals of European origin, like the Netherlands, predominantly express anabolism, which enable them to produce milk. These animals show a sulphuric constitutional basis with a degree of carbonism, which makes them slower and capable of maintaining calcium for a longer time in their deposits.

The Temperament of Diagnosis

This is a diagnosis of major importance to understand the morbid evolution. Temperament can describe the clinical prognosis and serve as a benchmark for aggravation or clinical rehabilitation.

The term used here refers to that presented by Allendy in his thesis, *the temperaments*, based on cellular metabolism and on the participation of endocrine glands in organism development. According to him the temperament diagnosis would show two metabolic benchmarks, the *tonicity* and *plasticity*. The *tonicity* would refer to the mechanisms of humoral response and regulation of circulatory and immune systems. The *plasticity* would be related to the ability of cell proliferation, tissue repair and organs development.

These metabolic benchmarks would present physiological changes over the individual life, assuming inherent features in each stage of organic development. According to the stage of life, the animal temperament would chronologically move, from childhood to old ages, in lymphatic (atoniplastic), blood (toniplastic), biliary (toniaplastic), and atrabiliar (atoniaplastic) stages. At each stage of development the organic disorders could modify the temperament by exhaustion of metabolic functions.

The lymphatic temperament would be considered peculiar to childhood, in which the metabolism is governed by the growth hormone. Thus, anabolism predominates, resulting in high *plasticity*; the immaturity of circulatory and immune systems causes lack of *tonicity*, that's why this phase is known as atoniplastic. The homeopathic medicines that seem to stimulate the commanded anabolism through growth hormone are mainly those called as constitutional medicines, such as

Calcarea carbonica, *Calcarea phosphorica*, and *Silicia* among others, such as *Sulphur*, *Natrum muriaticum* and *Barita carbonica*.

The blood temperament would correspond to the adolescence, the stage between childhood and adulthood. In animals it coincides with the emergence of the first rut in females and the reproduction ability in males. The animal organism replaces the growth hormone for the sex hormones, principally the 17-cetosteroids hormones, which complement the organic development and make emerge the secondary sexual characteristics. The anabolic steroids induced development results in animal *plasticity* and *tonicity*. The animal organism acquires great capacity of tissue regeneration and metabolic efficiency, in addition to the immune system maturation. This phase is called toniplastic because there is a large reactive and adaptive capacity, which makes the individual quite resistant to environmental adversities. The occurrence of castration, which is very common at this stage, both as in males as in females, can result in *tonicity* and *plasticity* injury, as well as predispose to certain organic imbalances. Hormonal imbalance can be related to febrile episodes, plethora and bleeding. The remedy that is compatible with this organic development phase is *Aconitum napellus*. Also, the use of contraceptive in small animals should badly interfere to the estrogen receptors function, generating ovarian dysfunction and bleeding, as well as predisposing to future mammary neoplasia. These clinical situations may require the use of biotherapics as *Oophorinum* and *Foliculinum*.

The biliary temperament is considered characteristic of those individuals in the adult life stage. Physiologically, the sex hormones no longer promote organic plasticity, when release of adrenal cortex hormones and endogenous cortical steroids are needed, to promote *tonicity* and a certain degree of cellular repair. The animal organism preserves its metabolism based on the humoral response capability, but suffers from organic overload, which damages the liver function and may result in different degrees of toxemia. It is necessary to keep good habits of life, avoiding the excesses and the inactivity to preserve the life quality. *Nux vomica* and *Lycopodium clavatum* are commonly used in these cases.

When the patient is old, the *plasticity* no longer exists since the biliary stage and the humoral response capacity is being lost, making the individual atoniaplastic. The atrabiliar temperament would translate the pattern response of elderly organism, in which the catabolic cellular reactions are widespread, harming the exonerative ability. Some feeble states could be related to the beginning of an "atrabiliar condition". In this cases the scleroses, the immunological difficulties and the detoxification deficiency appear. The homeopathic medicines toxicologically compatible to this phase are *Carbo vegetabilis* and the medicines from the *Ammonium* group.

The Diathesic Diagnosis

Diathesis (*miasma*) refers to the predisposition of contracting different diseases, from a similar nature and with hereditary character. The diathesis is not exactly the cause of disease, but a state of organic defense, an expression of organic dynamic adaptation to the environmental conditions. It can also be understood as the organism reactivity from the interaction between living organisms and the environment.

The identification harmful stimulus to which the organism is subjected is of great importance for the understanding of the homeostatic processes and also for the promotion of the healing process. In some life stages, where some physiological changes happen and after some pathologic states, such as malnutrition and severe infections, or when the organism is subjected to adverse environmental conditions for an extended period of time, a diathesic adaptation syndrome can be developed.

Classically, four different diatheses are described, psora, sycosis, syphilinism, and the tuberculinism (Carillo Júnior, 2000). These diatheses might occur alone or related, which is very common, making more complex the clinical condition.

It is accepted that organic overloads by mental or physical stresses, inactivity and poor nutrition might trigger off psoric reactivity, which are revealed to nerve excitability, hypertension, active congestion, and intense emunctories activity, affecting the skin and mucous membranes. At this stage there might be humoral *tonicity* response related to the immunoglobulin E (IgE) secretion with intense suprarenal participation, by synthesizing 17-cetosteroids. In a second stage, the exhaustion of 17-cetosteroids secretions, which is counterbalanced by the glucocorticoids secretion, might be related to an emerging clinical picture associated to passive visceral congestion and emunctories blocking. The need for catabolic or endogenous toxins elimination could promote increase of kidneys, intestine, skin, and some mucous functions, which can switch among them. The main clinical manifestations could be: increase in glandular secretions, dyspepsia and acceleration of peristalsis, bloating, spasms, diarrheas or constipations, type I allergic reaction to Gell and Coombs, and a tendency to infection by endoparasites and ectoparasites. The sulphuric biotype would present the most organic compatibility with psora, as those individuals that manifest blood or bile temperamental states.

The principal medicines used for the stage of psora are: *Sulphur, Calcarea carbonica, Nux vomica, Antimonium crudum* and *Ignatia amara*; while for the asthenia phase are: *Sulphur, Calcarea carbonica, Lycopodium clavatum, Graphites, Kali carbonicum, Ammonium carbonicum, Causticum, Psorinum,* and *Carbo vegetabilis.*

The sycosis diathesis may be considered as a syndrome of metabolic adaptation with specific forms of organic reactions that can differ from other diatheses. In sycosis there might be functional difficulty of phagocytosis, which takes to insufficient mucous defense mediated by IgA, predisposing the organism to infection, such as sinusitis, otitis, conjunctivitis, urethritis, and infectious gastroenteritis. These mucous would show increased production of mucopurulent secretions. Disarray in the sodium metabolism could be also seen, with consequent difficulty of mood mobility and formation of interstitial edema. Some patients might show neoplastic cell proliferations, with benign or malignant characteristics. The occurrence of disturbances in the sodium metabolism could make the patient more sensitive to humidity and predispose to edema formation, strokes and tissue infiltration. This clinical picture is called hydrogenism.

The clinical pictures showing IgA immunodeficiency point to anti-sycotic medicines that have tropism for skin or mucous membranes, such as *Thuya occidentalis*, *Pulsatilla nigricans*, *Sepia officinalis*, *Dulcamara*, *Hydrastis canadensis*, *Phytollaca decandra*, *Silicia* and *Rhus toxicodendrum*. In case of specific infections, some biotherapic medicines could be also used, such as *Colibacillinum*, *Enterococcinum*, *Pyrogenium*, *Streptococcinum*, *Staphylococcinum*, etc. The indication in cases of hydrogenism could be medicines of mineral (*Natrum*) origin: *Natrum sulphuricum*, *Natrum carbonicum*, *Natrum phosphoricum* and *Natrum muriaticum*. But some plant origin drugs, as *Thuya occidentalis* and *Apis mellifera*, also cover much of these cases.

Patients with cancer should be carefully evaluated under a surgical point of view and the tissue proliferation could be treated as a sycotic diathesis. Medicines that fit this physiopathological condition could be: *Thuya occidentalis*, *Conium maculatum*, *Nitric acidum*, etc. The therapeutic conducts in these cases might also include biotherapic medicines, such as DNA, RNA, Mioma, etc. Carbonic biotype individuals or mixed carbonism show diathesic compatibility with sycosis. The syphilis might be associated to irritating processes in vascular endothelium together with visceral phagocyte system compromising. This would entail some difficulty in oxygen and nutrients intake to the visceral parenchyma, compromising the intermediary cellular metabolism (glycides, lipids, proteins and electrolytes), together with weight loss. Viscera could be susceptible to infection or parasitic processes, sclerosis, cirrhoses and bleeding. In bones it would develop periostitis, osteomyelitis, cavities and ulcerations. In blood vessels, it would take to phlebitis, arthritis, strokes, embolism, aneurisms and varicose. Vegetations and inflammation would come up in the lymphoid system. These individuals commonly exhibit Type II hypersensitivity of resulted from interactions of IgG and IgM antibodies associated to complement activation. The toxicological picture of certain substances, such as *Argentum nitricum*, *Aurum metallicum*, *Bothrops lanceolatus*, *Lachesis trigonocephalus* and *Mercurius solubilis* shows identical pathophysiology to the syphilitic diathesic reactivity and could be indicated in these cases.

Phosphoric biotype animals have diathesic compatibility with syphilinism. When the syphilinism affects carbonic biotype animals there is a diathesic and constitutional incompatibility, which would make the prognosis reserved.

The tuberculinism could be linked to inherited genetic predisposition and/or acquired disorders related to hepatic phagocyte system that could be expressed by blood detoxification and mesenteric movement difficulty. Also, it could predispose to the occurrence of Type IV allergic manifestations. The compromised coloidogenic and proteinoclastic liver function would take to nonspecific reactions, as fever, ganglion swelling, bone demineralization accompanied by weight loss, dehydration, nervous hypersensitivity and constipation.

The main medicines to be used in these cases are from the phosphoric series and mineral origin as *Phosphorus*, *Calcarea phosphorica*, *Ferrum phosphoricum*, *Iodium*, *Arsenicum iodatum*, and *Sanicula aquae*. The biotherapic *Tuberculinum*, particularly the *Tuberculinum of Koch* are widely employed.

The diatheses may occur in isolated or related way, forming complexes, which make difficult to resolve diseases. The biotypologic and diathesic compatibilities should be examined aiming to the established the better therapeutic conduct. Generally, when the diathesis is treated first, it can be incompatible with the animal biotypology.

The Medical Diagnosis

The medical diagnosis is obtained from clinical, repertorial and pathogenetic images. Patient's clinical picture is built by all previous diagnostic conclusions. It portrays the structural and functional individual conditions and allows the understanding of their organic compatibilities and incompatibilities. All symptomatic totality of the patient must be transcribed into repertorial items to make the repertorisation. The resulting list of drugs composes the so-called *repertorial image* of the clinical case. The selected medicine will be that whose pathogenetic image – revealed by its mechanism of toxicological action – is similar to the physiophatological reactivity of the patient. Therefore, the medical diagnosis must be consistent to all analyzed aspects: clinical, clinical dynamic, biopathografic, biotypologic, temperamental and diathesic diagnoses.

Example of Clinical Case

Clinical Data

1. **Identification**: **Name**: Free **Species**: Canine **Race**: Beagle **Age**: 8 years **Gender**: Female
2. **Clinical history:**

Main complaint: It is becoming aggressive.

- **Current disease history**: it is becoming aggressive with the owner's daughter, even attacking her. In those occasions it likes to stay away from people, even to escape from home and disappears for 1–2 days. When it returns, it gives the impression that *nothing happened*, showing happiness but, once again, the aggression and sadness come back in a few moments. It seems that when the weather is warmer at home it expresses this behavior; with fresh weather, it seems more docile. It often shows vaginal purities. It has an intense appetite, it would eat until couldn't take anymore, eructing and flatting. It has a very fetid breath.
- **Physiological history**: it primarily eats feed and is gaining weight. It annually takes vaccine. No breeding.
- **Previous pathological history**: total hysterectomy due to pyometritis, diagnosed 4 years ago. It showed swelling in subsequent treatment with cortisone a year ago. It presents occasional claudicating.
- **Family history**: it has pedigree with a history background.

- **Clinical examination**: docile to the management, vital within the normal parameters, dry found, presence of peduncle warts on the face and lips, halitosis, presence of dental flame, body proportion, regular head and neck, cylindrical torso, costosternal angle greater than 90°, distended and tympanic abdomen, small vulva with pale mucosa, thin members disproportionate to the torso, and with regular angulations, painful sensation in the joints if the hand fingers if compressed.
- **Homeopathic clinical diagnosis:**

a) Clinical: interspecific aggressive behavior, arthrosis, papillomatosis and all recorded symptomatic.
b) Clinical dynamic: functional to light lesional.
c) Biopathografic: racial and management (weight, inactivity, diet).
d) Biotypologic: sulphurism, with carbonism acquired after castration.
e) Temperament: biliary.
f) Diathesic: Psora dominant (behavior, dyspepsia, dry skin, vaginal itching aggravating by heat) and sycosis (pyometra, edema and warts).
g) Repertorial image (repertorial items).

Number	Repertory section (SIHORE)20	Repertorial rubric
1	Mind	Irritability, Nervousness
2	Mind	Cholera
3	Mind	Aversion to company
4	Mind	Anxiety improves in outdoor
5	Generalities	Hot, at home it worsens
6	Extremities	Pain, gets better when walks
7	Stomach	Plenitude, after eating
8	Stomach	Belches
9	Mouth	Offensive smell
10	Skin	Drought
11	Skin	Flatus
12	Genital female	Purities

h) Repertorial image (drug selection).

Lycopodium	*Kali sulphuricum*	Sulphur	*Pulsatilla*	*Rhus toxicodendrum*
12*/31**	12/24	11/30	11/26	11/23
Calcarea carbonica	*Alumina*	*Carbo vegetabilis*	*Bryonia alba*	*Natrum muriaticum*
11/21	11/18	10/25	10/24	10/24
Nux vomica	*China officnalis*	*Arsenicum album*	*Carbo animalis*	*Kali carbonicum*
10/24	10/23	10/22	10/22	10/22
Sepia	*Graphites*	*Phosphorus*	*Lachesis trigonocephalus*	*Natrum sulphuricum*
10/22	10/21	10/20	10/18	10/18

* Number of symptoms covered by this medicine in material medica.

** Points computed from repertory.

i) **Medicated diagnosis:** *Kali sulphuricum* 6 pills 1x/day, for 15 days. Repeat after 15 days if necessary. After 45 days, a new evaluation is needed.
j) **Justification:** It was selected a product that presents correspondent pathogenetic image to the clinical picture of the patient. *Kali sulphuricum* is also compatible with the psora diathesis and sulphuric biotype of this patient, which is in a stage of structural change, by management failure and/or castration, acquiring adaptation carbonism, and also because it is under biliary temperament, with possible premature evolution for the atrabiliar.

Ambulatory Casuistic of the Veterinary Hospital – Homeopathy Unit of UFRRJ, Brazil

Homeopathy is a reality in the veterinary therapeutic field included in the curricular structure of Veterinary Medicine graduation program in UFRRJ with more than 15 years of practice routine. The discipline "Homeopathy" is under the universities criteria and regiments, offering educational, research and extension activities, with 50–60 graduation and post-graduation students per year, a defined target research and ambulatory attendance to different animal species.

The information about the patients attended during the year 2004 is described below, that showed the number of clinical cases and the species attended, the clinical and diagnostic indications, the homeopathic medicines prescribed and the number of consult returns.

The number of consults during 2004 was 325, being 26.7% related to returns. From this total, 79% were female and 21% male. This unequal distribution could be attributed to the obstetric surgical routine that is linked to Homeopathy discipline. A certain monthly oscillation could be attributed to interruptions in the academic calendar. Regarding to the different animal species attended, the prevalence of canine is evident (76%) which is in accordance with the casuistic situation of the veterinary hospital nowadays (Fig. 12.1).

The most important clinical situations verified and the more prescribed medicine during that period is exposed in Tables 12.1 and 12.2, respectively. The predominant surgical situations are because of the type of the hospital service, as cited above. So, the medicine most prescribed were those to treat the diseases classified as non-dynamic/mechanic nature and for reproductive organs.

Diseases of chronic and diathesic characteristic were also registered and its correspondent homeopathic prescriptions shows the presence of psora, sycosis and syphilinism. Among the medicines considered as constitutionals, those adapted to atrabiliar temperament and to punctual circumstances were used in a considerable frequency. The biotherapics, as the organotherapics, were also used frequently.

This statistic revealed the predominance of lesional clinical dynamic of patients from the Homeopathy Unit and characterized the homeopathic routine as a viable alternative in clinical cases of difficult resolution.

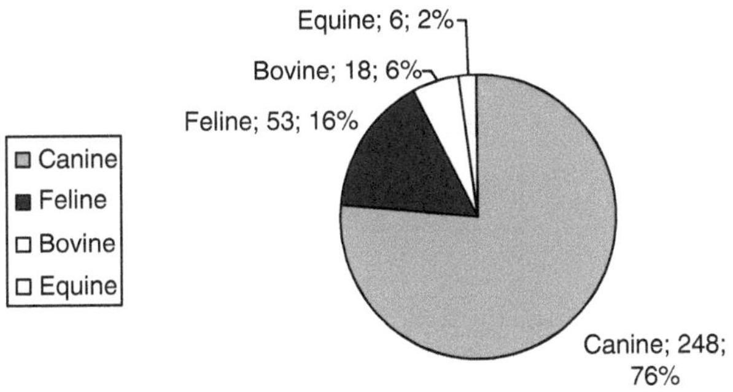

Fig. 12.1 Animal species attended and the respective frequencies in the Veterinary Hospital Homeopathic Service at UFRRJ, during 2004

Table 12.1 Clinical and diagnostic indications registered at the Veterinary Hospital Homeopathic service at UFRRJ, during 2004

Clinical and diagnostic indications	Number of cases	Frequency (%)
Pre and Pos-surgery	84	25.8
Dermatopathy	37	11.4
Chronic Cardiopathy	21	6.4
Mastitis	17	5.2
Metritis	14	4.3
Gastroenteritis	13	3.9
Behavior Disturbances	12	3.7
Pseudociesis	12	3.7
Hypothyroidism	12	3.7
Hip dysplasia	11	3.4
Chronic Constipation	8	2.4
Hyperadrenocorticism	7	2.1
Endoparasitism	7	2.1
Chronic Otitis	6	1.8
Salivary retention cyst	5	1.5
Bronco-pneumonia	5	1.5
Urolithiasis	5	1.5
Distocy	3	0.9
Ectoparasitosis	3	0.9
Epilepsy	3	0.9
Ehrlichiosis	3	0.9
Gynecologic hemorrhage	3	0.9
Toxic Hepatitis	3	0.9
Vaginal Prolapsed	3	0.9
Feline aestomatitis	2	0.6
Chronic Renal Insufficiency	2	0.6
Pathologic Puerperium	2	0.6
Venereal Tumor	2	0.6
Anemia	2	0.6
Dyslipidemia	1	0.3

(continued)

Table 12.1 (continued)

Clinical and diagnostic indications	Number of cases	Frequency (%)
Equine Pododermatitis	1	0.3
Orchitis	1	0.3
Others	15	4.7
Total	**325**	**100**

Table 12.2 Homeopathic Medicines prescribed at the Veterinary Hospital Homeopathic service at UFRRJ, during 2004

Medicines	Number of prescriptions	Frequency
Arnica montana	97	13.8
Calendula officinalis	84	12
Pulsatilla nigricans	57	8.2
Silicia	49	7
Biotherapy	45	6.5
Nux vomica	43	6.2
Calcarea phosphorica	35	5
Calcarea carbonica	27	3.9
Phosphorus	23	3.2
Phytolacca decandra	23	3.2
Sulphur	21	3
Conium maculatum	17	2.4
Thuya occidentalis	15	2.1
Lac caninum	12	1.7
Digitalis lanata	11	1.5
Crataegus oxyacantha	9	1.4
Gelsemium sempervirens	9	1.4
Lycopodium clavatum	9	1.4
Belladonna	8	1.2
Cactus grandifloris	7	1
Carbo vegetabilis	7	1
Mercurius solubilis	7	1
Apis mellifera	5	0.7
Sepia officinalis	5	0.7
Barita carbonica	5	0.7
Chelidonium	5	0.7
Drosera	5	0.7
Hydrastis canadensis	4	0.5
Sarsaparrilla	4	0.5
Calcarea fluorica	3	0.4
Cantharis vesicatoria	3	0.4
Berberis vulgaris	3	0.4
Bryonia alba	3	0.4
Carduus marianus	3	0.4
Hepar sulphur	3	0.4
Kali sulfuricum	3	0.4

(continued)

Table 12.2 (continued)

Medicines	Number of prescriptions	Frequency
Magnesia carbonica	3	0.4
Heckla lava	3	0.4
Plumbum	3	0.4
Symphytum officinalis	3	0.4
Kali carbonicum	2	0.3
Natrum muriaticum	2	0.3
Natrum sulphuricum	2	0.3
Actea racemosa	2	0.3
Rhus toxicodendrum	2	0.3
Secale cornutum	2	0.3
Causticum	2	0.3
Mezerium	2	0.3
Thlaspi bursa pastoris	2	0.3
Alumina	1	0.1
Hyoscyamus niger	1	0.1
X-ray	1	0.1
Nitric acidum	1	0.1
Sabina	1	0.1
Total	**704**	**100**

Conclusions

A common difficulty that exists in the practice of homeopathic therapy of veterinary medicine is the necessity to adapt the knowledge of homeopathic medical field – built on human proving and human clinical experience – to the veterinarian field. The simple transposition of human symptoms to different animal species may involve some equivocate applications of the *similia principle*, since the various animal species differ in several metabolic aspects.

Superior animals, belonging to fish, reptile, bird and mammal classes, show several physiological, metabolic and immune differences. This fact implicates in differences of toxic potential of certain substances, showing severe toxicity to some species but no reactivity at all for others. Prescribing homeopathic medicines by simple analogy to the human clinical picture may spread such errors and result in inefficacity of treatment.

The criterion for drug selection should be mainly based on the relationship between the substance toxicological picture and patient clinical and pathophysiological picture. The protocol presented here was built from the concepts developed by Carillo, Sihore and Pinto, in order to target this need. Thus, if there is high correlation between the disease pathophysiology and the toxicological picture of the prescribed substance, the probability of responsiveness increases. In short, the prescription is justified mainly by metabolic and pathophysiological references.

In the classical prescription of homeopathic drugs, there is the trouble of scale and pharmacological stimulation, which are variables without satisfactory and understandable theoretic basis. The proposed homeopathic clinical protocol establishes the use of 6CH, 12CH and 30CH, beginning by the lower to avoid the exhaustion of the action. The differentiation between mental or physical symptoms is not considered as a therapeutic benchmark because, in this protocol, both are considered belonging to the same hierarchy. However, the concept of unique medicine is maintained, except in cases where a particular prescription is needed, i.e., circumstantial or biotherapic indications.

References

Almeida BM, Pinto LF. Tratamento homeopático da Oesofagostomíase em bezerros no Estado do Mato Grosso. Homeopatia Brasileira; 2002, 8(1): 5–8.

Carillo Júnior R. Fundamentos de homeopatia constitucional. São Paulo: Ed. Santos; 1997.

Carillo Júnior R. Homeopatia, medicina interna e terapêutica. São Paulo: Ed. Santos; 2000.

Castilhos LR, Pinto LF, Almeida BM. Possibilidade terapêutica na linfadenite caseosa em caprinos no sistema orgânico de produção. Homeopatia Brasileira; 2002, 8(1): 55–59.

Castilhos LR, Souza JC, Pinto LF, Albuquerque FT, Filgueiras EP. Avaliação da terapêutica homeopática nas ovariopatias císticas de bovinos leiteiros. Homeopatia Brasileira; 2003, 9(1): 5–15.

Pfuetzenreiter MR. A epistemologia de Ludwik Fleck como referencial para a pesquisa no ensino na área de saúde. Ciência & Educação; 2002, 8(2): 147–159.

Pinto LF. Etologia medicamentosa. Homeopatia Brasileira; 1997, 3(3): 434–437.

Pinto LF. Rumos da homeopatia veterinária no Brasil. Homeopatia Brasileira; 1998a, 4(1): 505–506.

Pinto LF. Protocolo pós-operatório homeopático em pan-histerectomia de cadelas e gatas. Homeopatia Brasileira; 1998b, 4(1): 514–515.

Pinto LF. Ward casuistic of the veterinary homeopathy service in veterinary hospital at the Universidade Federal Rural do Rio de Janeiro. Homeopatia Brasileira; 1998c, 4(2): 517–523.

Pinto LF. Protocolo pós-operatório homeopático em mastectomia radical de cadelas e gatas. Homeopatia Brasileira; 2000, 6(1): 18–20.

Pinto LF. Imagem diatésica e biotipológica da Ehrlichiose canina – uma proposta terapêutica. Homeopatia Brasileira; 2001, 7(1): 10–12.

Pinto LF. Curso de formação de especialista em homeopatia para médicos veterinários – Apostila. Rio de Janeiro: Instituto Hahnemanniano do Brasil; 2006.

Pinto LF. O pensar complexo, os sistemas funcionais e a homeopatia; 2007 (in press).

Pinto LF, Almeida BM. O contexto da homeopatia na pecuária orgânica no Brasil. Homeopatia Brasileira; 2002a, 8(1): 23–28.

Pinto LF, Almeida BM. Terapêutica homeopática da papilomatose bovina em pecuária orgânica. Homeopatia Brasileira; 2002b, 8(1): 9–13.

Taravenic S, Pinto LF. Feline Lower Urinary Tract – A Study Case. Homeopatia Brasileira; 1998, 4(2): 567–575.

Thompson TDB. Can the caged bird sing? Reflections on the application of qualitative research methods to case study design in homeopathic medicine. BMC Medical Research Methodology; 2004, 4(4): http://www.biomedcentral.com/1471–2288/4/4.

Vijnovsky B. Tratado de matéria médica homeopática. Rio de Janeiro: Mukunda; 1980.

Epilogue

Chapter 13
Some Reflections About the Interest of Non-conventional Medicine in Public Health, Especially Homeopathy

Catherine Gaucher

Definitions

The use of non-conventional medicine in developing countries has been a challenge for many medical doctors during the last 20 years; among them, the "Homeopaths Without Borders" (HWB). Therefore, to think about the interest of this special form of medicine, called non-conventional medicine, in public health is inevitable.

One first shall give some prior definitions of the working field: what is non-conventional medicine and what is public health?

NON-CONVENTIONAL MEDICINE refers to the use of methods of treatment which have not been validated by the dominant scientific community.

PUBLIC HEALTH refers to the health condition of a community of human beings: in a town, in a country, on earth.

Historical Basis

Human being has been living on earth some millions of years ... facing, of course, some health problems. So, how did he manage them? Our modern medicine is a very little part, almost a single point, on the line of human's existence on earth. So, how did human beings some millions of years ago use to deal with health problems? That is the real question with non-conventional medicine and public health. It refers to the way by which millions of people, who do not have access to health services manage, deal with health problems they necessarily have to face.

NGO Homeopaths Without Borders, Paris, France. catherine450@gmail.com

Presented at the XXth Symposium of GIRI, São Paulo, 2006

What Is Traditional Medicine?

One shall refer to the definition of the World Health Organization (http://www.who.int/topics/traditional_medicine/en): "Traditional medicine refers to health practices, approaches, knowledge and beliefs incorporating plant, animal and mineral based medicines, spiritual therapies, manual techniques and exercises, applied singularly or in combination to treat, diagnose and prevent illnesses or maintain well-being".

Countries in Africa, Asia and Latin America use traditional medicine (TM) to help meet some of their primary health care needs. In Africa, up to 80% of the population uses traditional medicine for primary health care. In developed countries, adaptations of traditional medicine are termed "Complementary" or "Alternative" (CAM).

Increasing Use and Popularity

Traditional medicine has maintained its popularity in all regions of the developing world and its use is rapidly spreading in industrialized countries (http://www.who.int/topics/traditional_medicine/en). Examples:

- In China, traditional herbal preparations account for 30–50% of the total medicinal consumption.
- In Ghana, Mali, Nigeria and Zambia, the first line of treatment for 60% of children with high fever resulting from malaria is the use of herbal medicines at home.
- WHO estimates that in several African countries traditional birth attendants assist in the majority of births.
- In Europe, North America and other industrialized regions, over 50% of the population have used complementary or alternative medicine at least once.
- In San Francisco, London and South Africa, 75% of people living with HIV/AIDS use traditional medicine and complementary medicine.
- 70% of the population in Canada has used complementary medicine at least once.
- In Germany, 90% of the population has used a natural remedy at some point in their life. Between 1995 and 2000, the number of doctors who had undergone special training in natural remedy medicine had almost doubled to 10,800.
- In the United States, 158 million of the adult population use complementary medicines and, according to the USA Commission for Alternative and Complementary medicines, US$17 billion was spent on traditional remedies in 2000.
- In the United Kingdom, annual expenditure on alternative medicine is US$230 million.
- The global market for herbal medicines currently stands at over US$60 billion annually and is growing steadily.

Safety and Efficacy Issues

Unregulated or inappropriate use of traditional medicines and practices can have negative or dangerous effects. For instance, the herb "Ma Huang" (Ephedra) is traditionally used in China to treat respiratory congestion. In the United States, the herb was marketed as a dietary aid, whose over dosage led to at least a dozen deaths, heart attacks and strokes (www.who.int/mediacentre/factsheets/fs134/en). In Belgium, at least 70 people required renal transplant or dialysis for interstitial fibrosis of the kidney after taking a herbal preparation made from the wrong species of plant as slimming treatment (www.who.int/mediacentre/factsheets/fs134/en).

Because of that, WHO assesses that "scientific evidence from randomized clinical trials is only strong for many uses of acupuncture, some herbal medicines and for some of the manual therapies. Further research is needed to ascertain the efficacy and safety of several other practices and medicinal plants".

Biodiversity and Sustainability

In addition to patient safety issues, there is the risk that a growing herbal market and its great commercial benefit might pose a threat to biodiversity through the over harvesting of the raw material for herbal medicines and other natural health care products. These practices, if not controlled, may lead to the extinction of endangered species and the destruction of natural habitats and resources.

Another related issue is that at present, the requirements for protection provided under international standards for patent law and by most national conventional patent laws are inadequate to protect traditional knowledge and biodiversity.

A correct use and testing of homeopathic medicines should show its efficacy in many problems, without any secondary effects and with more respect of the biodiversity.

WHO Efforts in Promoting Safe, Effective and Affordable Traditional Medicine

The World Health Organization (WHO) launched its first ever comprehensive traditional medicine strategy in 2002. The strategy was designed to assist countries to develop national policies on the evaluation and regulation of Traditional medicine (TM) and Complementary and alternative medicine (CAM) practices and also to create a stronger evidence base on the safety, efficacy and quality of the TAM/CAM products and practices. Other strategies were also established: to ensure availability and affordability of TM/CAM including essential herbal medicines; to promote therapeutically sound use of TM/CAM by providers and consumers, and to document traditional medicines and remedies. At present, WHO is supporting clinical studies on antimalarials in three African countries; the studies are revealing good potential for herbal antimalarials. Other collaboration is taking place with Burkina Faso, the Democratic Republic of the Congo, Ghana, Mali, Nigeria, Kenya, Uganda, and Zimbabwe in the research and evaluation of herbal treatments for HIV/AIDS, malaria, sickle cell anemia and Diabetes Mellitus.

Face to this initiatives, is natural to suppose that WHO will also support homeopathic clinical trials in a near future, similar to those that HWB tried to realize in Peru for the cholera epidemics in 1991–1992.

Priorities for Promoting the Use of Traditional Medicines

Over one-third of the population in developing countries lack access to essential medicines. The provision of safe and effective TM/CAM therapies could become a critical tool to increase access to health care. While China, the Democratic People's Republic of Korea, the Republic of Korea and Vietnam have fully integrated traditional medicine into their health care systems, many countries are yet to collect and integrate standardized evidence on this type of health care. About 70 countries have a national regulation on herbal medicines but the legislative control of medicinal plants has not evolved around a structured model. This is because medicinal products or herbs are defined differently in different countries and diverse approaches have been adopted with regard to licensing, dispensing, manufacturing and trading.

The limited scientific evidences about TM/CAM's safety and efficacy, as well as other considerations, make it important for governments to formulate national policy and regulation for the proper use of TM/CAM and its integration into national health care, cultivating and conserving medicinal plants to ensure their sustainable use.

What Is Complementary and Alternative Medicine (CAM) and Non-conventional Medicine (NCM)?

As said in the beginning of this text, CAM refers to the use of methods of treatment which have not been validated by the dominant scientific community. According to the definition the WHO has given to that term, it should refer essentially to the use of traditional medicine in developed countries. But CAM includes more than the simple use of traditional medicine in developed countries, it represents a real challenge for developing countries.

NCM is the medical use of some of these methods of treatment that have been experimented and tested on a scientific way by professionals and health workers, from both occidental and oriental medicine. That is the reason why refer to "non conventional medicine" (NCM) is more appropriate than "complementary and alternative medicine" (CAM).

What are the advantages of NCM in public health?

NCM gets the advantages of CAM in most of the situations such as: the increasing use and popularity, not only in developing countries but in industrialized ones; the efficacy and safety have been proven in acupuncture, homeopathy and some manual therapies.

Regarding safety, plants are not always secure ... it requires a real scientific and botanical work which is not yet finished for the totality of them. Thus, the proposal for the use of NCM in developing countries regards essentially the use of homeopathic medicine as a complementary element in the totality of the CAM that already exists and can be used in the concerned area.

To treat tropical diseases, the proposal is to join four different ways:

- The use of the similarity law (classical homeopathy)
- The use of the typical French pluralistic vision of homeopathy and the treatment of the diathesis (or miasmas)
- The use of isotherapy (homeopathic preparation of the parasites or bacteria or viruses taken in their own environment, not isolated)
- The use of homeopathic remedies in order to diminish the toxicity of allopathic drugs (sometimes necessary)

One typical example is done, to illustrate this way of working with homeopathic medicine in order to improve results of medical care in public health.

One Typical Example: Bilharziosis

Bilharziosis is a well-known affection of the urinary tracts which is due to a parasite called Schistosoma. Three hundred millions of people in the world are concerned with the problem. The diagnosis is still essentially the clinical one; serology may confirm it rapidly.

First Step: What Does Similarity Law Offers to Us?

In the penetration phase, clinical symptoms are scarcely recognized and lead to inflammation remedies like Apis mellifica, Rhus toxicodendron and Sulphur. But one very interesting remedy, Antimonium tartaricum, has in its material medica the following characteristics: dysuria, stranguria, hematuria, albuminuria, inflammation of the bladder and of the urethra, burning sensation in the rectum, and emission of stools covered with blood and mucus.

The clinical complaints lead to the prescription of one of these remedies: Cantharis, Mercurius corrosives, Aconitum napellus, Cannabis sativa, Equisetum, Pareira brava, Terebenthina.

The cystoscopic aspect of the bladder mucosa injury looks like: Apis mellifica (edema), Cantharis (vesiculation), Hepar sulphur (hyper sensibility and bacterial infection with pus), Staphysagria (strawberries-like little tumors).

Both the clinical and the cystoscopic aspects refer to several homeopathic remedies, according to the similarity law of homeopathic medicine.

Second Step: What Does the Treatment of the Person Globality, Called Diathesis or Miasma Treatment, Offer to Us in that Particular Case?

Under the clinical point of view, four principal remedies could be proposed: Phosphorus (because of hemorrhages, burning sensation), Lycopodium (attempt of the urinary tract, aggravation between 5 and 7 p.m., does not like cold neither hot environment), Calcarea carbonica (wet cold sensation on the legs, sweats, acidity tendency, overweight), Staphysagria (bladder, pain when urinating, tumors on the genito-urinary mucosa).

The typical clinical symptoms are those of the diathesis called TUBERCULINISM and those of SYCOSIS.

Surprisingly, tuberculosis anatomo-pathological definition is that of a granuloma called "epithelioid granuloma" ... exactly the same kind of granuloma bilharziosis creates in the patient's bladder!

Third Step: The Use of an Isotherapic Remedy Made from the Parasite Taken in Its Own Environment

This has not been tried as far. Previous experiences had with malaria about that technique are promising. In this case, taking a part of the bladder's mucosa with the parasite inside (as for example taking one the little tumors by the time a cystoscopy is performed) and prepare a potentized remedy from it could give excellent results. To be tested....

Fourth Step: Prevention of Toxicity of Drugs

There are several drugs susceptible of killing the parasites, but with some side-effects as for example: vertigos (to which Bryonia, Cocculus indicus, Phosphorus, Conium maculatum could help), anorexia, loose of weight, anxiety with fear of death (to which Arsenicum album, Natrum muriaticum could be used) and pain in extremities (in this case Secale cornutum, Arnica montana, Natrum muriaticum could be indicated).

This way of working was already tried in several tropical diseases by HWB and a real efficacy was observed, as shown in some clinical research projects, listed bellow.

I – The example of MALARIA disease in Togo-Benin-Ghana:

The pilot study (realized in 1993, by Homeopathen Zonder Grenzen-HOLLAND) revealed, in 75 patients, 90.7% of amelioration.

The Double blind study (realized in 1993–94) revealed, in 74 patients (30 homeopathy; 25 chloroquine; 19 « out of study »), 72% ameliorated in the chloroquine group and 83.3% ameliorated in the homeopathy group.

II – The pilot study realized by HWB in Togo-Benin (1996):

The measure of the parasites density in blood smear after chloroquine compared to isotherapy of patient's blood revealed 50% of « zero parasites » at J3 in the isotherapy group against 12% in the chloroquine group of patients.

III – The example of CHOLERA disease and epidemics situation in Peru:

The pilot study (1991) showed in the 20 first cases (without discrimination) treated with eight principal remedies and rehydratation (iv): 5.51 perfused, instead of 10–151 usually, and 32 hours of hospital care, instead of the usual 5–7 days.

IV – The example of neonatology in Madagascar (1996):

Eighteen patients were included in the study, been nine boys and nine girls. The mother age was about 22–37 years (mean: 27 years), with or without previous pregnancies (12 patients first and 6 multiple). Sixteen cases were related to « risky births » for the newborns, being the Apgar Score between 3 and 10. In all cases, *Arnica Montana* was incorporated in the daily practice and the result was: 1.76% of weight loss in treated children and 3.26% of weight loss in the untreated ones.

Regarding to daily practice, some previous experiences in South America confirmed the suppositions. After the earth break in June 1994 in the Cordillera de los Andes, in Popayan (Colombia), a medical homeopathic Colombian doctor, Martha Emilia Parada Bernal managed to treat with homeopathy the whole population who had been placed under tents and all patients were cured in 2–48 hours.

Conclusion

Epidemiological protocols must be done, to demonstrate the usefulness of homeopathy in managing populations. One must be more audacious and take more time for that. It does not necessarily take such a lot of time, neither a lot of money.

INDEX

A
2-acetylaminofluorene (AAF), 85, 86, 88-91, 94
abiotic, 98, 101-104
acetylaminofluorene, 6
actinomycetes, 6
acute diseases, 187
acute inflammation, 7, 8, 85, 113-115, 118, 120, 121
adaptation, 20, 102, 103, 133, 190,191, 194, 198, 206
aerosols, 32
Africa, 206
agglomerates, 50
aggression, 188, 196
agitation, 4, 8, 11, 17, 29, 55, 137, 142, 144, 188, 189
agro-homeopathy, 107
allosteric protein, 127
Alternaria brassicicola, 106, 107
alternative, 11, 198, 206-209
ambiguity, 175
amount of information, 169
amphibian, 137, 138, 143, 145
amphibian metamorphosis, 8, 144
anabolism, 127, 192, 193
anger, 159-162, 166
anger with indignation, 157, 159, 174
anthroposophical medicine, 104
antigen, 8
anxiety of conscience, 159
Apgar Score, 211
apoptosis, 14, 129
aqueous solutions, 32, 33, 38, 43, 52, 53, 55, 58, 142
Aristotle, 50
arsenic trioxide, 103
arteries, 118, 119
artifacts, 5, 8, 12, 17, 32, 33, 72
Asia, 206
asthenia, 188, 189, 194
ATP, 128
atrabiliar temperament, 193, 198
attractor, 77, 78
autoimmune diseases, 14
autopoiesis, 4, 12, 17
Avogadro, A. 4
Avogadro's number, 4, 106, 130, 133

B
Bacon, Francis, 50
Bacon, Roger, 50
bacteria, 9, 17, 18, 33, 59, 99-101, 132, 209, 210
Balb/c mice, 16
barcode scanner, 138, 142, 144, 145
barley seeds, 100
basal membrane, 6, 92
basic research, 11, 83, 97, 98, 107
Basidiomycetes, 131
basophile degranulation, 5, 131, 132
basophiles, 5, 6, 70
Bastide, Madeleine, 3, 5, 6, 10-18, 20, 21, 32, 33, 132
Bayes, Thomas, 152
Bayes' formula, 153, 163, 178
Bayes'law, 161
Bayes'theorem, 151, 160, 162
Bayesian algorithm, 151
Bayesian statistics, 151, 152
behavior, 9, 10, 15, 16, 18-21, 31, 49, 50, 52, 54-61, 67, 75, 77, 113-123, 129, 140, 192, 196, 197, 199
behavioral disturbs, 9
behavioral patterns, 13, 14, 16
Benin, 211
Benveniste, Jacques, 5

Bergson, 22
Bernard, Henri, 191
bias, 69, 73, 155, 162, 173, 175, 177, 178
bilharziosis, 209, 210
biliary ducts, 84
biliary temperament, 192, 193, 198
bilirubin, 92
binomial chance calculation, 160
bioassays, 99, 102, 105
biochemical mechanisms, 6
biodiversity, 207
biological interaction, 33
biological meanings, 15
biological responses, 30-33, 76
biological specificity, 32
bioluminescence, 100, 101
biopathographic diagnosis/biopathographic diagnostic, 187, 188, 190
biophysical basis, 78
biosemiotics, 13
biotherapic(s), 193, 195, 198, 202
biotypes, sulfuric, phosphoric, carbonic, 191
biotypologic diagnosis/biotypologic diagnostic, 187, 188, 190, 191, 195, 196
blood brain barrier, 76
body expression, 11
Boenninghausen technique, 189
bold type, 157, 164
borderline cases, 175
botanical trials, 97
bound water, 51, 52
bovine pox-virus, 10, 11
bradikinin, 121, 122
Brazilian Homeopathic Pharmacopoeia, 58, 114
bubbles, 37, 43, 44
bulk, 58, 60
Burkina, Faso 208
bursin, 16
bursine, 6

C
cadmium, 7, 16, 102, 103
calorimetric properties, 30
CAM, 206-209
Canada, 206
canonical discriminant analysis (CDA), 152, 156-158, 167, 168, 170-172, 180
capacitance, 58
capacitor, 56, 59, 60
carbonic biotype, 195
carbonism, 191, 192, 195, 197, 198
carcinogen toxicity, 84
carrageenan, 7, 113-122
carrier, 15, 16, 20, 21, 32, 138
Cartesianism, 13
catabolism, 127
cattle, 107, 192
causality, 4, 11, 160, 176, 187
Cavendish, Henry, 50
cavitations, 43, 44
CD 203c, 5
cell organization, 7, 11, 17
cell phones, 8
cellular cycle, 83
cellular matrix interactions, 14
cellular system, 5
centesimal dilution, 8, 55
Central America, 107
central nervous system, 9, 84
chance, 106, 151, 153, 154, 160, 162, 163, 166, 180
chemical carcinogens, 84
chemical composition, 42, 50
chickens, 6
China, 197, 206-208
cholera, 197, 208, 211
chronic diseases, 187
circadian sleep cycle, 77
circannual rhythm, 17
circumstantial medicines, 189
Clarck's approach, 76
classical science, 18
clathrates, 31
cli-Fi-Col, 154, 155
clinic aggravations, 189
clinical diagnosis, 187, 188, 196, 197
clinical file, 154, 155
clinical history, 186-188, 196
clinical picture, 187, 188, 194-196, 198, 201
clinical practice, 9, 15, 42, 120, 154, 185, 191
clinical protocol, 185-202
cluster dynamics, 30
clusters, 33, 44, 50-53, 55, 56, 60
cohort study, 154, 156, 172
Colombia, 211
Committee for Methods and Validation (CMV), 154, 155
communication, 13, 14, 16, 18, 21, 8-0, 95, 185
comparable scale, 155
complementary, 29, 98, 206, 209
complex, 5, 12, 14, 16, 49, 56, 57, 69, 80, 92, 127, 131, 164, 166, 175, 194, 195
complexity, 13, 17, 32, 33, 50, 55, 77
complexity theory, 98
components, 56, 169
conceptual mistake, 31

conductivity meter, 38, 40
confirmation bias, 162, 177
connective tissue, 92, 113-123
consciousness, 20, 21
consensus procedures, 154, 172
constitutional medicines, 191, 192
constructive interference, 79
consultations, 158, 178, 179, 188
consumer health, 107
contaminants, 32, 42
context, 5, 6, 11, 13-17, 21, 70, 173, 185, 189
control group, 8, 9, 15, 16, 70, 71, 97, 143, 144, 178
co-occurrence, 175
correlation coefficient, 42
correlation matrices, 156-158, 166-168, 179
correlations, 30, 31, 38-40, 42, 44, 70, 156-158, 168, 179, 201
cress, 103
crystal, 53
curative action, 175, 176
curative potential, 175
cut-off values, 162, 165-167, 174
cybernetic regulation, 14, 21
cyclic or cascade reactions, 127
cyclines, 83
cyclones, 83
cyproheptadine, 7
cytokines, 14, 83, 84, 122
cytotoxic agents, 7

D

D_2O, 43, 54, 55
damped oscillator model, 80
data collection, 154, 155
databank, 154, 155
decision algorithm, 153
decision process, 153, 154
deeper organizational, 104
default value, 179
degranulated, 5, 116-122
degranulation, 5, 7, 122, 131, 132
delta band, 70-77
Democratic People's Republic of Korea, 208
Democratic Republic of the Congo, 208
density, 32, 50, 70, 72-78, 99, 142, 158, 211
density fluctuation, 32
Descartes, Rene, 13, 50
deuterium oxide, 31, *see also* D_2O
dexamethasone, 6, 8, 16, 85, 86, 88, 90, 91, 94, 95, 114, 117-123, 131
dexamethasone exerts, 85
dextrane, 7
diabetes mellitus, 208
diagnosis of temperament, 187, 188, 192, 196
diagnosis populations, 151
diagnostic test, 152
diarrhea from anticipation, 156
diasthesic incompatibility, 190
diasthetic diagnosis, 187, 193
diathesical chronic illness, 190
diathesical diseases, 188
diathesis, 188-190, 193-196, 198, 209, 210
dielectric constant, 50-57
diethylnitrosamine, 85, 89, 90
differential diagnosis, 167, 180
differentially diagnostic, 170
diffusion layer, 58
dinoflagellates, 100, 101
dipole moment, 50, 55, 56
discriminant analysis, 157, 172, 179, 180
discriminant function, 167, 170-172
dissipative structures, 31
dissolved gasses, 32
distilled water, 59, 140, 142
DNA synthesis, 83
DNA-viruses, 11
doctrine, 3, 4
dose, 7, 8, 54, 55, 67-72, 74-76, 78-80, 85, 86, 88, 98, 104, 107, 127, 130-133, 137, 142
dose-effect, 8, 68
dose-response, 67-70, 76, 79, 80
down-regulation, 10
drainage, 116, 189
drainers, 189
drug proving, 138
duality, 13, 133
duckweed, 99, 100
dwarf peas, 99, 100
dynamic effects, 20, 39
dynamic organization, 19
dynamical effect, 21, 31, 44
dynamical orientational complexity, 55
dynamic-clinical diagnosis, 187
dynamization, 4, 29, 31, 32, 37, 42, 132
dynamized systems, 4, 8, 29, 45

E

Earth, 49, 50, 205, 211
ECG, 69
edema, 7, 8, 113-116, 118-122, 194, 197, 210
EEG, 69, 70, 77-79
electrical conductivity, 38-40, 43, 44, 52, 56

electrical current, 78
electrical impedance, 56
electrical properties, 30
electrocardiogram, 69
electrochemical behavior, 57
electrochemisty, 56
electrode, 44, 57, 58, 70-72
electroencephalogram, 69
electromagnetic energy, 8, 143
electromagnetic fields, 16, 31, 32, 79, 137, 138, 140-145
electromagnetic nature, 32, 76
electromagnetic radiations, 8
electromagnetic signals, 77
electron microscopy, 129
electronic flow, 128
electrophoretic tests, 129
ELF EMF, 79
emunctories blocking, 189, 194
Endler, P.C., 7, 31, 99, 100, 131, 132, 137-140, 143, 145, 146
energy, 8, 31, 37, 44, 52-54, 60, 127-130, 143, 188
entangled states, 31, 33
environment, 4, 12, 17-20, 50, 69, 98, 106, 107, 114, 127, 129, 133, 138, 185, 191, 194, 209, 210
eosin, 93, 115, 116
epidermal growth factor (EGF), 84
epigenetics oscillations, 80
epistemological criterion, 186
equifinality, 78
equivalent circuit, 56, 57, 59, 60
Erlichia, 188
ethanol, 32, 42, 99, 102, 106, 142
ethical problems, 97
eukaryotes, 98
Europe, 206
ex vivo, 8
exothermic oxidation, 128
exotic characteristics, 49
experimental artifacts, 32, 33
experimental data, 4, 6, 12, 17, 21, 39, 41, 42, 44, 60, 61, 67, 76, 80
experimental design, 5, 15, 16, 98, 103, 114, 116, 140
experimental illustrations, 6, 17
experimental models, 5-8, 10, 12, 17, 84, 85, 120, 121
extremely diluted solutions, 44, 50, 58

F
family history, 190, 196
Faraday insulation, 71, 78
Fast Fourier Transform (FFT) algorithm, 73
fear of death, 156, 163, 165, 168, 170, 171, 174, 178, 180, 210
fear of the dark, 167
fever, 18, 188, 195, 206
feverfew, 102, 103
Fibonacci spiral, 130
fibronectin, 14
fluorism, 191
flux calorimetry, 44
footprints, 14
Fos, 83
fractal, 78
fractal concept, 78
free radicals, 128
free water, 51, 52
freezing point, 50
frequency, 56-61, 68, 71, 73, 79, 106, 143, 168, 189, 198-201
frogs, 137, 138
frond, 99, 100
fungal infections, 104
fungal pathogens, 98
fungi, 59, 98-101, 105, 106, 129-131

G
galvanic cell electrode potential, 44
γ-ray, 54, 55
GAP junctions, 95
gasification, 37, 43
gene expressions, 14
general organization, 12
generality, 175
Germany, 206
Ghana, 206, 208, 211
GHHOS scale, 156, 157, 167, 176
glasses type I, 41
glasses type III, 38, 40, 42
global attractor, 78
globality, 9, 11-13, 17, 159, 210
glucocorticoids, 122, 194
glucose, 128, 129
glycogen, 90, 92, 94, 95
glycolysis, 129
gold standard, 156, 172, 175-177
Gosner's stage, 139
grinding teeth, 156, 165, 170, 173
grooming, 9
Groupe International de Recherche sur l'Infinitésimal (GIRI), 5, 10, 118

H

Hahnemann, Samuel, 3, 29, 37, 79, 80, 152, 153
Hahnemann's disease classification, 187
Hahnemannian Centesimal Method, 38, 58, 86, 95
handmade, 37-48
heavy metals, 98
helenalin, 113, 122
Helmholtz layer, 58
hematoxylin, 92, 93, 115, 116
hepatectomy, 6, 83, 85, 87-91, 94
hepatic proliferation, 6, 85, 94
hepatic regeneration, 6, 16, 83-85, 87, 88
hepatocarcinogenesis, 16, 84, 85, 88, 95
hepatocellular carcinoma, 83
hepatocyte regeneration, 84
hepatocytes, 6, 83-85, 88, 90, 91, 95
herbal medicines, 206-208
herpes lips, 154, 156, 157, 162, 163, 165, 166, 168, 170, 173-175, 178 179
herpes virus, 10, 11
high dilutions, 4, 18, 21, 29, 31, 49-61, 71, 79, 85, 90, 95, 97, 132
high dilutions mechanisms of action, 97
Hill models, 67
histamine, 5-9, 67, 69-76, 80, 114, 121, 131, 132
histochemical analysis, 85, 94
histochemical characterization, 90
HIV/AIDS, 206, 208
Hoffmeyer, J., 13
holistic approach, 98
homeopathic algorithm, 178
homeopathic diagnosis, 153
homeopathic pharmacotechnics, 4, 7, 8
homeopathic phenomenon, 29-33, 44, 45, 186
homeopathic prescription, 151, 187
homeopathic symptoms, 151-181
homeopathic ultra-dilutions, 13
Homeopaths Without Borders (HWB), 205, 208, 211
homeostasis, 14, 69, 70, 77, 78, 133
homeostatic, 69, 70, 77, 194
hormesis, 33, 127-134
hormetic, 133
hormones, 13, 14, 99, 100, 137, 192, 193
host response, 70
humoral immune, 8
humoral immune reaction, 6, 8
hydrogen bond, 31, 32, 50-52, 54, 55, 58, 128
hyperstimulation, 138, 140, 143
hypothesis, 12, 13, 15, 16, 21, 22, 29-32, 38, 39, 43, 53, 54, 60, 80, 95, 98, 100, 122, 138, 178

I

idiosyncratic, 19
IL3, 16
image analysis, 92, 118
image of object, 21
image of patient and illness, 187
immunoglobulin, 8
immunoglobulin A (IgA), 194, 195
immunoglobulin E (IgE), 5, 194
immunoglobulin G (IgG), 5, 8, 195
immunoglobulin M (IgM), 8, 195
immunological studies, 8
immunosuppression, 16
impedance, 56-61
impedance analysis, 56-58, 61
impedance diagram, 57
impendance spectroscopy, 56-58, 60
in vitro studies, 7
incidence, 95, 178, 179
independent entities, 166
indication for the medicine, 157, 164, 165, 173
indignation, 157, 159, 166, 174
individuality, 14, 15, 20
individualization, 15
inflammatory cells, 8, 115, 116
inflammatory mediators, 5, 7
inflammatory process, 6, 113, 120, 121, 189
information, 12-18, 20, 21, 30-33, 38, 44, 51, 56, 60, 61, 76, 78, 80, 85, 103, 104, 122, 127, 128, 130, 132, 138, 145, 154, 164, 169, 170, 180, 185, 198
information carrier, 15
information exchange, 15, 17
information matrix, 15
informational models, 33
informative event, 19, 20
informative relationship, 20
initiators, 6, 84, 85, 88, 89
interaction between symptoms, 166
interface effects, 58, 60
inter-group variance, 158
interpretative hypothesis, 12, 21
inter-relatedness of symptoms, 156, 158
interrelations, 164
interrelations of symptoms, 172, 179
intracellular chain-reactions, 13
intra-group variance, 158
ionic contaminants, 32
irritability, 160, 188, 197

iso-endopathy, 7, 8, 11, 137
iso-endopathy principle, 70
isopathic, 7, 15, 95, 102, 104, 106, 107, 131, 133
isopathic ultra-dilutions, 13
isopathy, 11, 15, 16, 33, 86, 90, 107, 133
isotherapic, 7, 15, 17, 210
italics type, 157,165
itch-behavior, 9, 10
itching, 188, 197
Ito cells, 84

J
journal *Nature,* 5
Jun, 83

K
Kent, James T., 18, 176, 188
Kenya, 208
keynote, 180
Krebs cycle, 128, 129
Kupffer cells, 90, 95

L
laboratory animals, 9, 70
laboratory plants, 9
Lagache, Agnès, 3, 12, 13, 15-18, 32
laminin, 14, 84
language, 20, 175
Latin America, 206
Les Entretiens Internationaux de Monaco, 5
leukocytes, 116, 117, 121, 122
level of communication, 13
LiCL lithium chloride, 31, 3, 54, 55, 58-61
life quality, 188, 189, 193
light lesional stage, 189
likelihood ratio (LR), 152-169, 172-181
linear fitting, 40, 41, 43
linear primary modulation, 13
liquid water, 50, 52, 53
liver, 83-85, 87, 88, 90, 94, 95, 189, 195
living beings, 4, 12-14, 18-20, 69, 70, 77, 79, 80
living systems, 7, 11, 12, 14-17, 33, 79, 129
loquacity, 156, 164, 165, 168-170, 173, 175, 177
low potencies, 43, 103, 104
LR assessment, 154, 156, 162, 165, 167, 173, 177, 179
LR project, 154, 173, 178, 179
luminescence, 32, 53, 54
luminescence methods, 129
luminometry, 130
lupus, 16
lymphatic absorption, 115, 116, 120, 122
lymphatic vessels, 116, 119, 122
lymphocytes, 14, 115, 117, 121

M
Madagascar, 211
malaria, 206, 208, 210, 221
Mali, 206, 208
margin for error, 175
Masson's trichrome, 90, 91, 93, 94
mast cell histamine degranulation, 7
mast cells, 6, 116-122
materia medica, 15, 29, 151, 160, 162, 171-173, 175, 180
Materia Medica Validation, 154, 155, 160, 163, 167, 173
Mathematical, 16, 38, 39, 42, 57, 76, 77, 158
mathematical analysis, 69, 71, 72
mathematical models, 9, 12, 67, 79
Maturana, H.R., 14, 99
MCH, 14, 99
meaningful, 12, 15, 19, 20
mechanical, 4, 19
mechanical agitation, 55
mechanical arm, 59
mechanical machine, 38
mechanical succession, 37, 38, 43
mechanicism, 4
mechanism, 3, 4, 6, 7, 9, 12, 31, 37, 44, 51, 55, 68-70, 79, 80, 83, 85, 97, 121, 122, 127, 128, 130, 133, 134, 185, 187, 190, 192, 196
mechanism of action, 4, 7, 85, 185
mechanistic, 4, 10, 68
mechanistic paradigm, 3, 13
medical diagnosis, 196
medicine population, 151, 157, 162, 168, 180
melting point, 50
memory, 8, 11, 17, 29, 30, 133, 152
memory of water, 5
metabolic oscillations, 80
metabolism, 95, 127-129, 191-195
metaloproteins, 128
metamorphosis, 7, 8, 131, 137, 138, 140, 144
methyl-histamine, 5
miasma, 193, 209, 210
micro domains, 51, 52
microbial trials, 97
microorganisms, 18, 97-108, 133
microwave oven, 8, 138, 142, 144, 145

middle ages, 4, 113
migration, 8, 85, 116, 117, 121, 122, 191
minimum syndrome of maximum value, 187
misleading entries, 163, 180
misleading repertory-entry, 163
mitochondria, 128, 129
mobile phone, 138, 142, 144, 145
modeling, 51, 56, 57, 67
modified Glick, 90
modulation, 5, 12, 13, 79, 80, 94, 121, 122
modulator, 5-8, 11, 17, 70, 120
molecular communication, 14
molecular interactions, 52
molecules, 5, 7, 9, 12-14, 30-32, 42-44, 49-56, 92, 106, 128
molecule-receptor system, 9
monisms, 13
monocytes, 115, 117
monomer, 56
Montpellier, 5
morbid process, 188, 189
morphometric analysis, 87, 89
morphometrical characterization, 90
Mpemba, Erasto, 50
multi-centre trial, 100, 108
multicentric study, 5, 7
multidisciplinary, 33
multivariate analysis, 156, 158, 167, 172, 179
Myc, 83

N
NaCl (sodium chloride), 31, 43, 54, 55, 100, 102, 103
nanobubbles, 31
nano-pharmacology, 134
natural sciences, 13, 67
nematodes, 98, 105, 106
neonatology, 211
neuron, 7, 78
NΓk(kappa)B, 83, 85
nidogen, 14
Nigeria, 206, 208
NMR, 32
no linearly network, 12
Nocardia, 129-131
noise, 32, 79, 80, 104
non linear system, 58, 79, 80
non local effects, 31
non-conventional medicine (NCM), 205-211
non-linearity, 8, 9, 11, 12, 17, 67-81
non-molecular information, 12, 14, 16
non-monotonic dependence, 53, 54
non-REM sleep, 70
normality, 10, 11, 17, 73
North America, 206
noxae, 104
NREM sleep, 77
nuclear transcription, 83
Nyquist diagram, 56
Nyquist theorem, 71

O
occurrence, 58, 151, 152, 154, 162, 173, 175, 178, 188, 191, 193-195
oncogenesis, 84
open cohort studies, 154, 156, 172
organic agriculture, 107
organizational structure, 49, 51, 53
Organon, 3, 29
organotherapics, 189, 198
oscillating, 80
oscillations, 16, 17, 76, 80, 132, 133, 145, 198
oscillator, 76, 80, 129
oscillatory, 5, 8, 68, 75, 76, 133
oscillatory pattern, 80
oscillatory potency-effect curve, 5, 8
oval cells, 6, 87, 95
oxydoreductases, 128, 129, 131, 132
oxygenases, 128, 129

P
p53, 83
painful feeling, 188
paradigm of the sign, 13
pathogenetic images, 187, 196, 198
pathogenetic symptom, 159, 160
pathological conditions, 6, 13, 100-102, 106, 122
pathological history, 188, 190, 196
pathological patterns, 9
pathophysiological moment, 188, 189
pattern recognition, 158, 164
patterns of symptoms, 179
peculiar symptoms, 152, 153, 162, 181
periodic acid-Schiff (PAS), 92, 116, 118
peroxidase, 120-132
personal traits, 151
Peru, 208, 211
pharmacogenetics, 151
pharmacology, 6, 9, 12
Pharmacopoeia, 38, 42, 98
phenolic, 129, 130
phenomenological approach, 12
phenomenological models, 33
phenotype, 6, 16, 89, 90

phosphoric biotype, 188, 195
phosphorism, 191, 192
phosphorus, 15, 100, 106, 153, 168, 188, 191, 195, 197, 200, 210
physical chemistry properties, 38, 43
physical properties, 31, 33, 50, 52
physical-chemical models, 29-31
physical-chemical variations, 30
physical-chemistry parameters, 69
phytopathological models, 98, 104
Phytophthora infestans, 106, 107
piece of information, 14, 15, 20, 21
Pillars, 4
Pivot tables, 157
placebo, 9, 97
plain type, 157
plant- and microorganism-based experimentation, 97
plant hormones, 99, 100
plant immunity, 133
plants, 9, 32, 97-108, 127, 131-133, 207, 209
plasticity, 192, 193
pleural cavity, 116
PMN (polymorphonuclear) cells, 115, 117, 121, 122
Poisson regression, 67
polytraumatic injuries, 113
positive gradient, 155
possessiveness, 188
posterior odds, 153
post-surgical edema, 121
potency-effect curve, 5, 8, 9
potentization, 4, 33, 37, 38, 43, 44, 103, 130, 131
Potentization *see also* dynamization
potentized (or dynamized) systems, 4, 8, 12, 29-33
potentized gamma radiation, 103
power laws, 77
practice experience, 172, 180
preclinical studies, 97
preneoplasic nodes, 6, 16
pre-neoplastic lesions (PNLs), 89-91, 94, 95
prescriptions, 151, 154, 157
prevalence, 77, 151-157, 160-163, 165, 168, 173, 174, 178, 179, 181, 198
previous conditions, 7, 11, 12, 17
Prigogine, I., 31
principal component analysis (PCA), 156-158, 167-169, 179
promoters, 84, 89
prospective assessment, 152, 156, 162, 173, 177
prospective investigation, 161, 162, 173
prospective LR research, 160, 172, 174, 176, 178, 180
prostaglandins, 7, 114, 121, 122
proto-oncogenes, 6
proving, 9, 32, 38, 122, 138, 155, 159, 201
Prussian blue, 90, 93-95
psora, 194, 197, 198
psychological disturbs, 9
public health, 205-211
pyruvic acid, 128

Q
qualitative, 80, 90, 175
quality of life, 9
quantitative, 76, 90, 175
quantum nature, 33
questionnaire, 180
quiet wakefulness, 77

R
RADAR®, 154, 157
RADAR-Synthesis®, 157
radiation, 8, 52, 103, 133, 145
Raman, 32
Rana catesbeiana, 7
Rana temporaria, 7, 137, 139, 140
random disturbance, 69
rare symptoms, 153
recall bias, 162, 178
receiver, 15, 16, 20
recursive, 12, 67
red light scan, 138, 142, 144, 145
reference "gold" standard, 172, 175-177
refraction, 52
regenerative hepatic process, 83
relative growth rate, 75, 76
repertorisations, 152, 154, 166, 175, 196
repertory, 151, 152, 154, 155, 157, 158, 160, 162-167, 172-175, 177, 179-181, 197
reportorial images, 196, 197
reproducibility, 60, 97, 103, 104, 138, 173
Republic of Korea, 208
resistance, 41, 42, 56, 58, 60, 70, 84, 105, 106
resistant hepatocyte model, 88, 91
resonance, 20, 21, 79, 133
resonant behaviors, 60
respiratory chain system, 129
retrospective analysis, 152, 161, 181
retrospective assessment, 173
retrospective investigation, 173, 178
retrospective study, 154, 155, 174
Robinson Crusoe, 14

rubrics, 151, 152, 155, 157, 159, 160, 162-164, 166, 173-175, 177, 181, 197

S

Saccharomyzes cerevisiae, 101
satellite pictures, 158
scale, 55, 77, 78, 155, 156, 176, 201
Schizosaccharomyces pombe, 101
scree-plot, 169
self response, 79, 80
Self-Organized Critically (SOC), 77, 78
self-similarity, 77
semantic, 9, 13
semantic objects, 12, 13, 15, 17
semantically similar, 155, 173
semantics, 166, 173
semiosis, 4
semiotic notions, 12
semiotics, 13, 29
sensibility, 31, 32, 210
sensitive systems, 12, 13, 104
sensitivity to injustice, 154, 156, 157, 161, 164, 166, 168, 170, 171, 173, 177
serotonin, 121, 122
severe lesional stages, 189
side effects, 9
signals, 20, 32, 54, 58, 69, 71, 77, 79, 80, 127, 145, 188, 189
signal-to-noise ratio (SNR), 32, 79
silica, 31, 32, 42
silver nitrate, 100
silver staining, 118
similarities, 8-11, 15, 16, 21, 29, 30, 33, 77, 79, 165, 170, 173, 187
similarity law, 160, 209, 210
similarity models, 8
similia principle,
similitude, 33, 122
sinusoidal curve, 132
sleep, 69, 70, 72, 76, 77, 80, 156, 160, 164, 165, 173
sleep pattern, 70
sleep-wake, 70
slow water dynamics, 30
slow-wave activity, 70
soft science, 13
software requirements, 178
solvent, 15, 16, 30, 32, 44, 55, 71, 76, 142, 145
solvent structure, 32, 45
Sorites paradox, 175
Sorites vagueness, 175
South America, 107, 111
specific heath, 50
specificity, 5, 7, 11, 15, 17, 18, 20, 32, 61, 92, 131, 191
spectral density, 70, 72-74
spectrophotometry, 115, 130
spectroscopy, 56, 58, 60, 129
SPSS, 158
stable cells, 83
stationary state, 77
statistical knowledge, 180
statistical uncertainty, 173, 174
statistical variance, 152
stem cells, 6, 84
stochastic, 69, 173
stochastic resonance, 79
stressed organisms, 104
structural closure, 17
structural models, 45, 51
structure, 17-21, 30-33, 49-54, 60, 67, 69, 79, 84, 87, 90, 93, 97, 98, 100, 118, 128, 129, 191, 198
succession, 11
sulphuric constitution, 191
sunflower, 103
surface tension, 50
sustainability, 207
sustainable agriculture, 98, 107
sycosis, 194, 195, 197, 198, 210
symbolic information, 13
symptoms, 4, 7, 9-11, 17, 19, 20, 29-31, 80, 104, 105, 107, 122, 151-181, 185-188, 196, 197, 201, 202, 209, 210
synergic action, 13
syphilinism, 194, 195, 198
system sensitivity, 104

T

TCR, 14
temperamental diagnosis, 188
tension, 50, 56, 58
theorem, 71, 152, 160, 162
theoretical explanations, 30, 145
theoretical models, 31, 33, 97
Theory of (the) Corporeal Signifiers, 3-22
thermoluminescence, 31, 43, 54, 55
thymulin, 16, 21
thymus, 14
thyroxin, 7, 8, 131, 137, 138, 140-145
T-lymphocytes, 14
TM/CAM, 207, 208
tobacco, 98, 105, 106
tobacco mosaic virus (TMV), 105, 106
Togo, 211
toluidine blue, 116-118

tonicity, 192-194
totality, 9, 11, 17-19, 187, 196, 209
toxicology, 7, 187
traditional medicine, 206-208
transforming growth factor (TGF), 84
triggering, 5, 8, 127
tuberculinism, 194, 195, 210
tuberculosis, 210
tumoral lineage, 7
turbulence, 43, 44
type I allergic reaction, 194
type I hypersensitiveness, 9
type II hypersensitivity, 195
type IV allergic manifestations, 195
type of shaking, 131

U
Uganda, 208
ultra highly diluted substances, 97
ultra-diluted dexamethasone, 6, 8
ultra-dilutions, 3-22, 67-81
ultra-high dilutions, 43, 50, 52-56, 58, 60, 83-95, 98
ultra-low dose, 68, 72-76, 79, 80
United Kingdom, 206
United States, 38, 206, 207
unsuccussed medium, 100, 101

V
Vaccinia, 10
Vaccinotoxinum, 11
vague symptoms, 154, 162, 166
vagueness, 157, 162, 172, 175, 177
validation criteria, 29, 30
Van der Waals interactions, 52
varicela, 10
vascular permeability, 114, 116, 121, 122
vasodilatation, 9
very low doses, 43, 127-133
veterinary homeopathy, 186
veterinary medical specialty, 186
vibration modes, 51
Vietnam, 208
vincristine, 43
viral pathogens, 98
virology, 10
viruses, 10, 11, 97-108, 209
viscosity, 50
vital functions, 104
vitalism, 4, 21
voltage fields, 78
vortex, 8, 37, 43, 44

W
water, 5, 7, 15, 16, 29, 30, 32, 38-45, 48-61, 71, 86, 92-94, 102, 103, 105, 106, 114, 116, 128, 133, 137, 139-145
water cluster hypothesis, 32
water dynamics, 30, 32
water infrared absorption spectra, 51
water memory, 29, 30, 133
wave graph, 132
wave length, 46, 53, 132
wave shape, 131, 133
waves, 53, 54, 70, 77, 132, 133, 145
weakening of quantum theory, 31
weakness, 188, 189
wheat, 98, 102, 103, 105
wheat seeds, 103
Winchip® program, 154
World Health Organization (WHO), 206-208

X
x-ray, 55, 138, 142, 144, 145, 201

Y
yeasts, 98, 102

Z
Zimbabwe, 208
zona, 10, 11

FSC
www.fsc.org
MIX
Papier | Fördert
gute Waldnutzung
FSC® C083411

Kolibri 360 GmbH
Ferdinand-Jühlke-Straße 7
99095 Erfurt, Deutschland
produktsicherheit@kolibri360.de